Women in Industry 4.0: Shaping World Foundations in Engineering, Business, and Sustainability

Series Editors: Tilottama Singh, Richa Goel, and Kayla L. Tennin

According to a World Economic Forum report, only 22% of AI professionals globally are female. This series will bring stories straight from the heart of these innovative women who have established themselves, to inspire millions and create a wave of impact. This will be a series of books and authors that will focus on sustainability through technology and entrepreneurship in Industry 4.0. The series will act as a global platform for bringing together organizations and women sharing their life cases, research, and grounded theory which will be an interesting read for engineers, businesses, government, researchers, and students working on startups and contributing to gender diversity.

Each book in the series is dedicated to illuminating women's research and achievements in key, targeted areas of the contemporary tech world and raising awareness of the pivotal work women are undertaking in areas of importance to our global community toward sustainable development. It will be a step in bringing forth the reality of what overshadows women in workplaces and the strategic attempts of organizations to pave the path for women's representation in companies and fields of engineering.

Books that focus on women in engineering and technology exploring and describing how women have greatly contributed to the world at large through these fields and their many subsections is a focal point of this series.

If you are interested in writing or editing a book for the series or would like more information, please contact Cindy Carelli, cindy.carelli@taylorandfrancis.com.

Women-Led Innovation and Entrepreneurship: Case Studies of Sustainable Ventures and Practices
Edited by Esra Sipahi Döngül, Serfie Uguz Arsu, Richa Goel, and Tilottama Singh

African Women Trailblazers in STEM: Pioneering Technological and Economic Advancement
Edited by Uzo Mkparu, Richa Goel, Tilottama Singh and Jade Mkparu

Women, Tech, and Enterprise: A Path to Sustainable Equality in South Asia
Edited by Kavita Indapurkar, Neeru Sidana, Richa Goel, and Tilottama Singh

Tech Startups Driving Innovation: Empowering Women in Bangladesh
Edited by Quazi Tafsirul Islam, Richa Goel, and Tilottama Singh

Tech Startups Driving Innovation

Empowering Women in Bangladesh

Edited by
Quazi Tafsirul Islam, Richa Goel, and
Tilottama Singh

CRC Press
Taylor & Francis Group
Boca Raton London New York

CRC Press is an imprint of the
Taylor & Francis Group, an **informa** business

Designed cover image: Shutterstock

First edition published 2027
by CRC Press
2385 NW Executive Center Drive, Suite 320, Boca Raton FL 33431

and by CRC Press
4 Park Square, Milton Park, Abingdon, Oxon, OX14 4RN

CRC Press is an imprint of Taylor & Francis Group, LLC

ISBN: 978-1-032-89759-2 (hbk)
ISBN: 978-1-032-89838-4 (pbk)
ISBN: 978-1-003-54483-8 (ebk)

DOI: 10.1201/9781003544838

Typeset in Times
by Newgen Publishing UK

Contents

Preface

The global tech industry has seen tremendous innovation, disruption, and growth over the past few decades. Yet, despite this rapid evolution, women remain significantly underrepresented, particularly in tech startups across emerging economies. Bangladesh, with its burgeoning entrepreneurial ecosystem and dynamic digital economy, presents a unique landscape to explore how women are navigating, shaping, and transforming the tech startup sector.

Tech Startups Driving Innovation: Empowering Women in Bangladesh was born out of the need to document, understand, and celebrate the stories, challenges, and achievements of these pioneering women. This book brings together comprehensive research, case studies, and interviews to shed light on the experiences of female tech entrepreneurs in Bangladesh — their motivations, barriers, strategies, and contributions.

As authors, we believe that sharing these narratives is vital not only for academic inquiry but also for inspiring policy reforms, educational initiatives, and support systems that promote greater gender diversity in technology. We hope this work will serve as a foundational reference for scholars, policymakers, investors, and aspiring entrepreneurs who seek to foster an inclusive startup ecosystem in Bangladesh and similar emerging markets.

We extend our sincere gratitude to the women who generously shared their journeys with us, and to Taylor & Francis for their support in bringing this book to life. It is our hope that *Tech Startups Driving Innovation: Empowering Women in Bangladesh* will spark meaningful conversations and drive impactful change toward a more equitable future in technology and innovation.

During the preparation of this book, the editors and contributing authors made limited and responsible use of generative artificial intelligence (AI) tools across the manuscript. Such tools were used to varying degrees in most chapters for language refinement, editorial workflow support, reference organization, and indexing preparation. In addition, AI-assisted literature review and research support tools, including ChatGPT (OpenAI, GPT-4 series/web version available during manuscript preparation), Grammarly Premium (current web version during manuscript preparation), Consensus (current web version during manuscript preparation), Elicit (current web version during manuscript preparation), and Research Rabbit (current web version during manuscript preparation), were particularly utilized in Chapters 1, 2, 3, 5, and 6 to support literature discovery, thematic organization, evidence synthesis, source mapping, referencing processes, and idea development.

All AI-assisted outputs were carefully reviewed, validated, and substantially revised by the authors and editors to ensure accuracy, originality, and scholarly integrity. These tools were used solely to enhance efficiency and clarity, and not to replace core intellectual contributions, critical analysis, interpretation, or final scholarly judgment.

No generative AI tools were used to create or manipulate original research data, figures, tables, or images included in this book. All interpretations, analyses, and conclusions presented in this work remain the sole responsibility of the human authors.

The use of AI tools in this book complies with the Taylor & Francis AI Policy and reflects a commitment to transparency, accountability, and ethical scholarly practice.

April 2025
Dhaka, Bangladesh

About the Editors

Quazi Tafsirul Islam is a faculty member at North South University, Bangladesh, and the Founder of Bengal Consulting Group (BCG). Professor Islam is an acclaimed academic, AI educator, and management consultant with over a decade of experience spanning higher education, HR consulting, and organizational development. He is widely recognized for integrating sustainability, innovation, and AI in business education and strategy. Professor Islam has authored six academic and professional books and over 100 business case studies, and his work spans human resource management, organizational behavior, sustainable business models, and digital transformation.

Richa Goel is an accomplished academic with 23+ years of experience in economics and management. She is an associate professor at Symbiosis Centre for Management Studies, Noida, Symbiosis International University, PUNE. A Gold Medalist in Economics, she holds a PhD in management with expertise in diversity management. She has to her credit more than 20 books and numerous papers published with leading publishers. She is involved with training and has conducted many international and global conferences and has been a trainer for many FDPs concerning SDGs.

Tilottama Singh is a certified human resources analyst from IIM and a proficient academic, researcher, and trainer with more than 14 years of experience in the field of human resources and work dynamics. She is presently the Head of Department in the School of Management, Uttaranchal Institute of Management, Uttaranchal University, Dehradun. Her areas of expertise are human resources, entrepreneurship, and strategy, and her teaching concentration includes human resources, strategy, and law. She started her career in the hotel industry with Leela Kempinski in the corporate head office in human resources and training. Later, she joined the education sector with the topmost universities IMS Unison Group and Amity University, the leading university in Pan Asia. She has obtained her master's in human resources and management with distinction. Before that, she graduated in economics with honors and law with a distinction, with an added vocational course in mass communication and analytic certification from IIM. Having an enriched research portfolio with Scopus and international refereed journals, book chapters, and conference presentations, she has been awarded the Best Presentation Award at international conferences. She also serves as a member of AIMA and acts to lead and liaise between the student community and the industry delegates, with a keen interest in training and building sustainable models for business and society.

Contributors

Jashim Uddin Ahmed
North South University
Bangladesh

K. S. Amritha
Department of Psychology
The American College
Madurai, India

Anta Atalantia
North South University
Bangladesh

Sabina Begum
Department of Analytical and Applied Economics
Utkal University
India

Alok Ranjan Behera
Department of Analytical and Applied Economics
Utkal University
India

B. Daniel
Department of Psychology
The American College
Madurai, India

Anisur R. Faroque
University of Vaasa
Finland

Tilka Farzana
Department of Marketing and International Business
North South University
Bangladesh

Zulfiquer Ali Haider
DAN Department of Management & Organizational Studies
Western University
Canada

Faiz Ibne Hossain
North South University
Bangladesh

Mohammed Imran
School of Women's Studies
Utkal University
Odisha, India

Zulkarin Jahangir
North South University
Bangladesh

Effat Binte Kabir
Independent University
Bangladesh

Nabila Kamal
Independent University
Bangladesh

Saslina Kamaruddin
Faculty of Management & Economics
Sultan Idris Education University
Malaysia

Zarjina Tarana Khalil
Department of Marketing and International Business
North South University
Bangladesh

Md. Rahat Khan
Institute of Business Administration (IBA)
University of Dhaka (Savar)
Bangladesh

Samiha Mamun
Institute of Business Administration (IBA)
University of Dhaka (Savar)
Bangladesh

Samuel Mursalin
North South University
Bangladesh

Farzana Nahid
North South University
Bangladesh

Rifah Nanjiba
North South University
Bangladesh

Mostafizur Rahman
North South University
Bangladesh

Rakshanda Samin
We-Fi at the World Bank
Bangladesh

D. Simona
Department of Psychology
The American College
Madurai, India

Amreen Talukdar
North South University
Bangladesh

Apoorva Thakur
MKA Legal
India

Rani Thanacoody
The University of Nottingham
United Kingdom

1 Empowering Women in Bangladesh's Tech Startup Revolution

A Framework for Participation, Innovation, and Growth

Anisur R. Faroque, Faiz Ibne Hossain, Samuel Mursalin, Zulkarin Jahangir, and Quazi Tafsirul Islam

1.1 INTRODUCTION

In the heart of South Asia, Bangladesh is quietly undergoing a revolution. Bangladesh's participation in international trade with textile and agricultural products has been undoubtedly extraordinary. Recently, the country has been rapidly emerging as a breeding ground for technological innovation and entrepreneurship.

While still in its early stages, Bangladesh's startup ecosystem is reshaping its economic and social landscape.

Interestingly, the role of women in business and technology sets a powerful precedent for gender equality in the digital age.

Women's participation in startups and access to venture capital remain glaringly limited worldwide. According to research by Startup Genome, only 20% of startups had at least one female founder in 2019, a figure that has doubled since 2009 but still shows significant underrepresentation (Stefanuto, 2023).

In terms of funding, the disparity is even more startling. According to TechCrunch, all-women founding teams received only 1.9% of all venture capital funding in recent years (Davis, 2023). According to a Boston Consulting Group (BCG) study, Europe had only 10% women-led startups, and a minuscule 2% raised venture capital successfully (Lauprete, 2023).

These statistics underscore the uphill challenges women face in the startup ecosystem and also highlight the need for continued efforts to promote gender equality in startups.

Over the past decade, Bangladesh has witnessed admirable growth in its startup scene. From health tech companies like Praava Health to travel tech companies like

DOI: 10.1201/9781003544838-1

ShareTrip, the country's entrepreneurs are leveraging technology to solve pressing issues.

Bangladesh has caught the attention of numerous internationally recognized investors, such as SoftBank, Valar Ventures, and Prosus Ventures. According to LightCastle Partners, over 90% of investment in Bangladesh's startup ecosystem came from international investors, demonstrating global optimism and commitment to the Bangladeshi startup ecosystem.

However, the true potential of this tech revolution lies not just in the innovations themselves but in who's driving them forward. Women are generally underrepresented worldwide in startups. The increasing participation of women entrepreneurs in Bangladesh serves as a catalyst for short- and long-term economic growth and social progress.

While the strides toward women's participation in Bangladesh's startup ecosystem are admirable, barriers still persist. Access to specialized education, venture capital, societal expectations, work-life balance, and discrimination still pose major hurdles.

Addressing these challenges is crucial for the overall success and sustainability of Bangladesh's startup ecosystem.

This chapter aims to explore the multifaceted landscape of women's participation in Bangladesh's tech startup revolution. We will examine initiatives in education, entrepreneurship programs, and policy support that are paving the way for increased female participation in Bangladesh's startup ecosystem.

Furthermore, we'll analyze the tangible impact women are making on the startup scene. We'll showcase success stories and explore how recognition of these achievements is elevating Bangladesh's profile on the global stage.

Next, we investigate how women-led startups are contributing to job creation, driving GDP growth, and attracting both domestic and international investments. These economic outcomes are closely related to broader social changes, as successful women entrepreneurs become role models and challenge stereotypes to empower others.

Through a detailed case study of ShareTrip, we provide insights into the challenges and triumphs of female entrepreneurship in Bangladesh's tech sector. Sadia Haque's example offers valuable lessons to numerous young women who consider entrepreneurship as a career choice.

Finally, we'll conclude with a set of strategic recommendations to empower women in the startup space.

Evidently, women's empowerment in Bangladesh's startup ecosystem is more than just gender equality. It is a foundation of the country's economic, social, and inclusive future.

1.2 ENTRY POINTS FOR WOMEN IN TECH-BASED STARTUPS

As Bangladesh's startup ecosystem continues to flourish, it's crucial to ensure that women have ample opportunities to participate and lead in this dynamic sector. Empowering women in tech-based startups fosters gender equality and drives innovation, economic growth, and social progress.

1.2.1 Education and Skill Development Programs

The journey of empowering women in Bangladesh's tech startup ecosystem begins with foundational education and skill development. Several initiatives have been launched to promote STEM (Science, Technology, Engineering, and Mathematics) education among girls from an early age to ensure women are equipped with the necessary competencies.

Organizations such as BRAC and the Jaago Foundation have pioneered programs that encourage young women to pursue STEM education. These initiatives include scholarships, mentorship opportunities, and exposure to real-world STEM applications. By creating an inclusive learning environment, these programs aim to break traditional gender norms and inspire more women to embark on tech-related careers (Jaago Foundation, n.d.).

Moreover, specialized training programs focusing on digital skills are crucial. Institutions like Bangladesh Women in Technology (BWIT) offer courses in coding, data analysis, and digital marketing, tailored specifically for women. These programs provide not only technical skills but also soft skills such as leadership and entrepreneurship, preparing women to thrive in the competitive startup landscape (Bangladesh Women in Technology (BWIT), n.d.).

The Women ICT Freelancers Entrepreneurs Development Program is a collaborative initiative by BWIT, Aspire to Innovation (A2i) under the Information, Communication, and Technology (ICT) Division of the Bangladeshi government, and the SME Foundation. Funded by the Ministry of Women Affairs, this program aims to develop female ICT freelancers and entrepreneurs, particularly targeting underprivileged rural women (Bangladesh Women in Technology (BWIT), n.d.).

The program offers training in various skills, such as graphic design, data entry, and accounting, with the goal of enabling women to generate income from home (Bangladesh Women in Technology (BWIT), n.d.).

It has successfully conducted numerous training sessions across all 64 districts of Bangladesh, providing essential resources and support to help women sustain their businesses in a competitive market (Bangladesh Women in Technology (BWIT), n.d.).

The Women and ICT Frontier Initiative (WIFI), launched in 2016, is a flagship program by the United Nations Asian and Pacific Training Centre for Information and Communication Technology for Development (UN-APCICT), in collaboration with BWIT. WIFI focuses on promoting women's entrepreneurship in Asia and the Pacific through ICT capacity development (APCICT, n.d.).

The initiative aims to empower women economically by enhancing their knowledge and skills in business management and ICT usage. It also supports government leaders and policymakers in creating an enabling environment for women entrepreneurs through gender-responsive policies and programs (APCICT, n.d.).

WIFI has trained thousands of women, helping them leverage digital technologies to grow their enterprises and navigate challenges in the business landscape (APCICT, n.d.).

Additionally, corporate partnerships play a significant role in skill development. Companies like Microsoft and Google have launched initiatives to train women in various tech domains in Bangladesh. These programs often come with certifications that enhance employability and career advancement prospects. By fostering a culture of continuous learning, these initiatives empower women to stay ahead in the fast-paced world of technology.

Microsoft's 'Code Without Barriers' initiative aims to address the gender gap in the tech industry by providing training, networking opportunities, and mentorship to female developers and tech entrepreneurs. Launched across nine countries, including Bangladesh, this program focuses on the cloud computing, artificial intelligence (AI), and digital technology sectors. By fostering a culture of continuous learning and offering certifications, it empowers women to stay competitive and advance in their tech careers (Mathew, 2021)

Similarly, Google's Women Techmakers Program offers visibility, community, and resources for women in technology. Members can access events, workshops, and mentorship opportunities, helping them build a supportive network and enhance their skills. The program also encourages women to become ambassadors, leading and organizing tech events in their communities. This initiative not only boosts the presence of women in tech but also inspires them to take on leadership roles and make a significant impact in the industry (Google Women Techmakers, n.d.).

Another crucial aspect is the creation of safe and supportive learning environments in formal and informal educational institutions. Community centers and co-working spaces dedicated to women provide an inclusive atmosphere where they can learn, collaborate, and innovate without societal pressures or biases. These spaces often host workshops, hackathons, and networking events that further enhance their learning experience and professional growth.

1.2.2 Entrepreneurship Programs

Entrepreneurship programs play a pivotal role for women aspiring to launch their own tech startups. These programs provide a holistic approach by offering mentorship, networking opportunities, and access to funding.

The ICT Division's 'ShePower' initiative stands out as a beacon of support for women entrepreneurs in Bangladesh. This program offers mentorship from industry veterans, helping women navigate the challenges of starting and scaling a tech business. Additionally, it facilitates networking events where women can connect with potential investors, partners, and fellow entrepreneurs (The Financial Express, 2018).

Funding remains a significant barrier for women-led startups. To address this, several organizations have launched women-centric funding programs. For instance, Biniyog Briddhi provides grants and low-interest loans to women entrepreneurs. These programs enable women to focus on innovation and business development by alleviating financial constraints (Biniyog Bridhhi, n.d.).

Bangladesh Women Investors Network (BWIN) is a pioneering initiative launched by the Bangladesh Angels Network (BAN) to support women entrepreneurs and investors. As a sister chapter of BAN, BWIN focuses on filling the early-stage financing gap for women-led and women-focused startups by providing investment, mentorship, and access to networks.

The network aims to create a more diverse equity investor pool and promote gender-lens investing, ensuring that promising women-led ventures receive the funding they deserve. BWIN also offers high-quality mentoring and strategic support to its portfolio companies, helping them grow and succeed in the competitive startup ecosystem (Dhaka Tribune, 2022).

Entrepreneurship bootcamps and hackathons specifically designed for women create an immersive learning environment. These events foster collaboration, ideation, and practical problem-solving, equipping women with the confidence and skills needed to succeed in the startup world. Platforms like Women in Digital Bangladesh host regular events that bring together aspiring women entrepreneurs to brainstorm and develop innovative solutions.

Global networks such as Women's Startup Lab provide a broader platform for Bangladeshi women entrepreneurs that they can use to gain international exposure and access to global markets. These networks offer mentorship, investment opportunities, and knowledge-sharing sessions, empowering women to scale their startups beyond national borders.

Women's Startup Lab is a global initiative that provides Bangladeshi women entrepreneurs with access to international markets and exposure to Silicon Valley's innovative ecosystem. Founded by Ari Horie, this program offers a unique residential experience in Silicon Valley, where participants can immerse themselves in the local startup culture, gain insights from top mentors and advisors, and build a strong network of global connections.

Through various programs, including the WiSE24 global event, Women's Startup Lab helps women entrepreneurs develop their leadership skills, boost their confidence, and accelerate their business growth on a global scale (Women's Startup Lab, n.d.).

Collaborations with universities and research institutions also play a vital role in promoting entrepreneurship among women. Academic programs that combine technology and business education prepare women to innovate and lead in the startup ecosystem. Partnerships with industry leaders ensure that these programs remain relevant and aligned with the latest market trends and technological advancements.

1.3 POLICY SUPPORT

Government and non-governmental organizations play a crucial role in fostering an enabling environment for women entrepreneurs through policy support. The Bangladesh government's ICT Policy includes provisions aimed at reducing gender disparities in the tech sector.

Non-governmental organizations (NGOs) such as the Bangladesh Women Chamber of Commerce and Industry (BWCCI) advocate for policies that support women in tech. These organizations work closely with the government to implement programs that address the unique challenges faced by women entrepreneurs.

Partnerships between the government, NGOs, and the private sector are essential for creating a comprehensive support system.

Startup Bangladesh Limited is the flagship venture capital fund of the ICT Division, sponsored by the Government of Bangladesh. Launched in March 2020 with an allocated capital of BDT 500 crores, this initiative aims to foster innovation, create new jobs, and develop technical skills. Startup Bangladesh supports technology-based innovations and promotes underrepresented tech groups by providing investment in equity, convertible debt, and grants to pre-seed, seed, and growth-stage startups. The program also offers in-kind support to startups and stakeholders, contributing to the growth of a vibrant startup ecosystem in Bangladesh (Startup Bangladesh Limited, n.d.).

Initiatives like Startup Bangladesh provide resources, mentorship, and funding opportunities for women-led startups, fostering a vibrant ecosystem where women can thrive.

Furthermore, international collaborations and funding from global organizations like the World Bank and UN Women have bolstered policy support for women in tech. These collaborations bring in global best practices and resources, enhancing the effectiveness of local policies and programs. For instance, the World Bank's Women Entrepreneurs Finance Initiative (We-Fi) provides financial and technical assistance to women entrepreneurs in Bangladesh, helping them overcome barriers and scale their businesses.

Additionally, policy frameworks that promote gender equality in corporate governance and leadership roles can significantly impact the startup ecosystem.

Advocacy for inclusive work environments is another critical aspect of policy support. Bangladesh can pave the way for increased female participation in its tech startup revolution. Empowering women in this sector is more than a matter of gender equality; it is a critical driver of innovation, economic growth, and social progress.

1.3.1 Impact on the Startup Scene

Women's participation in Bangladesh's startup scene is a transformative force reshaping the entire landscape. As women entrepreneurs increasingly take the helm of innovative ventures, they bring with them fresh perspectives, novel solutions, and a keen understanding of underserved markets.

The impact extends beyond individual success stories. From fostering diversity and inclusion to garnering international recognition, women-led startups play a pivotal role in elevating Bangladesh's global startup profile.

The multifaceted impact of women's participation in the tech startup scene is driving innovation, enhancing ecosystem resilience, and positioning Bangladesh as an emerging hub for tech entrepreneurship on the world stage.

1.3.2 Innovative Solutions

Women-led startups in Bangladesh are increasingly at the forefront of introducing innovative solutions. Women entrepreneurs disrupted multiple established industries, such as logistics, health, financial technology, travel, and education, which have long dominated the business landscape.

1.3.2.1 Foodpanda Bangladesh

While Foodpanda is a global brand, its operations in Bangladesh have been significantly shaped by women in leadership roles. Ambareen Reza, the co-founder and managing director, has been instrumental in driving Foodpanda's growth in the country since 2013. Under her guidance, Foodpanda Bangladesh has expanded beyond food delivery to include deliveries of groceries, medicine, and other essentials.

Ambareen's leadership has been particularly noteworthy during the COVID-19 pandemic, when she steered the company toward supporting restaurants and customers during challenging times. Foodpanda's success has paved the way for other women in Bangladesh's tech and logistics sectors.

1.3.2.2 Shohoz

Founded by Maliha Quadir in 2014, Shohoz has evolved into a comprehensive online ticket-booking platform that offers bus, ship, and train tickets. Maliha's vision and leadership have been pivotal for Shohoz's growth, making it one of Bangladesh's most prominent tech startups.

Her background in investment banking and experience with international tech companies have been crucial in attracting significant foreign investment to Shohoz. Maliha's success has made her a role model for women entrepreneurs aspiring to thrive in a traditionally male-dominated industry.

1.3.2.3 Praava Health

Sylvana Q. Sinha founded Praava Health in 2018 with a mission to transform healthcare in Bangladesh. As a Bangladeshi-American lawyer and development professional, Sinha brought a unique perspective to healthcare innovation. Under her leadership, Praava has introduced a patient-centric model that combines telemedicine with brick-and-mortar clinics.

1.3.2.4 ShareTrip

ShareTrip has rapidly become Bangladesh's leading online travel agency. Sadia Haque is the visionary co-founder and CEO of the company. ShareTrip's core differentiator has been a solid understanding of Bangladeshi people's need to explore top local and international travel destinations within affordable budgets.

Her leadership has been crucial in navigating the challenges posed by the COVID-19 pandemic to the travel industry and demonstrating resilience and adaptability.

1.3.2.5 Arogga

A relatively younger company than the previous ones but doing exceptional work, Arogga is an online pharmacy and healthcare platform. Co-founded by Rosina Mazumder, Arogga's vision is to make healthcare more accessible and affordable in Bangladesh.

Under Mazumder's leadership, Arogga has streamlined medication delivery and introduced features like online doctor consultations, addressing critical healthcare needs, especially for women who may face barriers in accessing traditional healthcare services.

Above are just a few from a long list of exceptional women-led tech companies in Bangladesh. Numerous other women-led companies, such as SBK Tech Ventures and The Legal Circle, provide phenomenal services to help other technology companies thrive.

1.3.3 Diversity and Inclusion

The increasing participation of women in Bangladesh's startup ecosystem is an instigator of enhanced creativity, innovation, and resilience. Diversity and inclusion in the startup scene bring a wealth of perspectives, experiences, and problem-solving approaches crucial for addressing complex societal challenges and driving sustainable growth.

Research consistently shows that diverse teams outperform homogeneous ones. A study by BCG found that companies with more diverse management teams have 19% higher innovation revenues. This finding is particularly relevant in the context of startups, where innovation is the lifeblood of success (Lorenzo et al., 2018).

Bijon Islam, co-founder of LightCastle Partners, notes that women-led startups in Bangladesh are more likely to identify and address gaps in products and services that cater to women consumers, a market segment often overlooked by male entrepreneurs (Islam, 2024).

Moreover, diverse startup ecosystems tend to be more attractive to international investors and partners. Tina Jabeen, former CEO of Startup Bangladesh, notes that the increasing visibility of women entrepreneurs in Bangladesh's tech scene has played a significant role in attracting foreign investment and partnerships (Seshasayee, 2023).

1.3.4 Recognition and Awards

Recognizing women entrepreneurs through awards and accolades plays a crucial role in shaping Bangladesh's global startup profile. These acknowledgments not only celebrate individual achievements but also inspire future generations and draw international attention to the country's vibrant startup ecosystem.

Internationally, Bangladeshi women entrepreneurs are making their mark. Bibi Russell, founder of the sustainable fashion tech startup Bibi Productions, was honored with the UNESCO Artist for Peace award in 2021, highlighting the global potential of Bangladesh's women-led startups.

The Inspiring Women in Tech Awards, launched in 2020 by the ICT Division of Bangladesh, have become a significant platform for recognizing women's contributions to the tech industry.

Moreover, these awards are shaping Bangladesh's global startup profile. The success stories of women entrepreneurs like Sadia Haque, Ivy Huq Russell of Maya, and Sylvana Q. Sinha of Praava Health have been featured in international media outlets like TechCrunch and Forbes, putting Bangladesh's startup scene on the global map.

Sonia Bashir Kabir, founder and managing general partner of SBK Tech Ventures; Rubaba Dowla, founder of Pulse Healthcare Services; Tasfia Tasbin, co-founder and CEO of Markopolo AI; Jahnnobi Rahman, co-founder and CEO of Relaxy; Saraban Tahura, founder of Turtle Ventures Ltd.; Rizvana Hredita, co-founder and COO of Hydroquo+; Shomy Chowdhury, co-founder of Awareness 360; and many other Bangladeshi women entrepreneurs have won prestigious international awards and represented Bangladesh on the global stage.

As Bangladesh continues to nurture its startup ecosystem, the recognition of women entrepreneurs will play a pivotal role in attracting talent, investment, and global partnerships. These success stories and accolades are not just celebrations of individual achievement—they are powerful catalysts for change, inspiring a new generation of women to enter the tech startup space and contribute to Bangladesh's digital future.

1.4 SHARETRIP'S JOURNEY

The founding journey of ShareTrip was born out of a shared passion for travel and a vision to transform the industry in Bangladesh. Early on, Sadia Haque and her husband Kashef Rahman recognized a significant gap in the market—there was no single, simple, one-stop travel solution. As avid travellers who loved exploring new countries on a tight budget, they understood first-hand the challenges that came with planning trips and realized there was a huge opportunity in digitizing and simplifying the process for others.

The duo's love for travel inspired them to take a leap of faith, leaving behind their respective corporate jobs to pursue a dream of making travel more affordable and accessible. Under the banner of Travel Booking BD, they started small, providing offline services with just a handful of employees, but their passion resonated with others. While operating in the local market, they began bootstrapping on the backend technology by taking baby steps in developing their online platform. Their first investors saw their vision for a more transparent and inclusive travel experience in Bangladesh and that they are here not for the short term but, rather, to be part of the marathon for this industry, and the investors' support allowed them to scale their idea.

In 2018, they rebranded their entire service portfolio as ShareTrip and transitioned into a fully digital platform, becoming Bangladesh's foremost online travel platform and the only true online travel agency (OTA), providing end-to-end travel solutions for the end users. This move opened up endless possibilities, allowing them to provide a comprehensive range of services beyond bookings, from hotel reservations to holiday packages. They built ShareTrip with the goal of making travel stress-free for everyone, staying true to the principle of working with passion and a long-term vision.

According to Ms. Haque, their journey has been one of innovation and perseverance, and they remain committed to reshaping the travel landscape in Bangladesh, pushing the boundaries of what's possible in the travel-tech space (S. Haque, personal communication, October 21, 2024).

1.4.1 Impact on the Tech Startup Ecosystem

According to Sadia Haque, ShareTrip is a platform dedicated to women's empowerment and transformation and also provides them with the tools and opportunities they need to thrive. By nurturing these bright minds, we're not only fostering a future rich with diverse perspectives but also breaking new ground in our society.

From the very beginning, ShareTrip aimed to evolve the traditional travel market into a fast-paced online travel service. The intent came with a strong ideation of the ways to build an army as part of the ShareTrip organization. According to Sadia Haque, having had previous corporate careers, the core principles of running and leading an organization were already known to them. Ms. Haque's aim was to establish the trust of the people and to make them see the benefits of having a local online travel partner that understands and responds to each of their needs. Keeping that in mind, ShareTrip's strategic vision centered around making travel solutions more accessible through technology. They introduced significant innovations, such as their mobile app and user-friendly website, which provided seamless travel planning for customers. This focus ensured competitive pricing and enhanced convenience, even as ShareTrip integrated with offline agents to cater to diverse customer needs. Ms. Haque stated how proud she has been as, with the help of her team and her experience, ShareTrip has established itself as a key contributor in shaping the Bangladeshi travel industry.

As a woman leading a tech company, Sadia Haque feels a strong responsibility to inspire and support more women to enter and thrive in the tech space. The tech industry thrives on diversity. ShareTrip values the unique perspective of strong, resourceful young women in STEM fields. She is fostering a future brimming with diverse perspectives and groundbreaking ideas! She wants women in ShareTrip to feel empowered to grow, take on leadership roles, and make their mark in the tech industry. The company believes in building a diverse and equitable tech industry, and they are committed to leading the way by showing that women are and will keep playing a pivotal role in shaping the future of technology in Bangladesh.

According to Sadia Haque, leading a tech startup brings fresh challenges every day but also incredible opportunities to innovate and make a difference. That is why she is passionate about supporting initiatives empowering young women in STEM fields. According to her, these programs equip them with the skills and resources they need to thrive.

Furthermore, she strives to play her organization's part in encouraging young Bangladeshi women to venture into the tech startup space by making it an example of a community that enables and supports them in every step of their journey. As part of this vision, she is personally engaged in various communities that are in favor of mentorship, ideation and initiation of new business entities, or working opportunities for women in the tech space. These are significant positive responses from the community and from society on a larger scale that will assist us all in bringing more women into this field (S. Haque, personal communication, October 21, 2024).

1.4.2 Lessons and Strategic Insights

Sadia Haque firmly believes technology has been at the core of solving many of the challenges in various sectors, including the travel sector, especially in a fast-evolving market like Bangladesh. From making travel more accessible to ensuring a seamless customer experience, ShareTrip used technology as a powerful tool to reshape how people plan and book their trips. Their dedicated tech team has developed bespoke solutions that cater to the specific needs of the Bangladeshi market and its customers. This understanding of local preferences has enabled ShareTrip to offer tailored services that resonate with travelers.

One of the most impactful ways they have leveraged technology is through their mobile app and website. Traditionally, booking flights or hotels involved a lot of hassle—visiting travel agents, long waiting times, and limited information. With the company's digital service platforms, they have simplified the entire process. This convenience has been a game-changer for travelers, making ShareTrip the go-to platform for many.

Sadia Haque's biggest advice to startups looking to incorporate AI, mobile apps, and digital solutions is to always start with a clear understanding of their customers' needs. Technology for the sake of technology rarely works. If startups can identify where their customers are experiencing pain points or where they can offer something faster, easier, or more personalized, then the digital solutions they implement will truly add value.

Furthermore, when it comes to AI, it's not just about automating processes but about enhancing the customer experience. AI can be used to predict user behavior, personalize recommendations, and improve customer service. For other digital solutions, it's important to remain agile and adaptable. The digital landscape is evolving rapidly, so startups need to stay ahead by adopting emerging technologies and keeping a close eye on market trends. But it is not recommended to try to implement everything at once—start with what aligns most closely with one's business goals and scale as needed. Sadia always emphasizes customer focus, simplicity in design, and making data-driven decisions, all of which are critical for startups looking to stay competitive in a digital-first world (S. Haque, personal communication, October 21, 2024).

1.5 STRATEGIC IMPLICATION

Women-led startups in Bangladesh, as in many parts of the world, face significant barriers in accessing venture capital. Research highlights that women-led ventures receive significantly less funding than their male-led counterparts, with gender biases influencing investment decisions (Guzmán & Kacperczyk, 2018). Addressing this gap requires targeted interventions to improve access to funding for women entrepreneurs.

Governments and private-sector investors should establish venture capital funds dedicated to women-led startups. International models such as Golden Seeds and Astia Angels have demonstrated success in increasing women's access to early-stage capital by creating gender-lens investment opportunities (Robb, Coleman, & Stangler, 2014). Bangladesh can replicate these models to bridge the funding gap.

Networks such as the BAN can incorporate gender-focused initiatives by promoting women investors and funding women-led startups. Research suggests that increasing female representation among investors leads to higher funding rates for women entrepreneurs (Solal, 2019).

Special financial instruments such as low-interest loans, grants, and microfinance targeted at women entrepreneurs can ease the capital constraints. Similar initiatives have been successful in narrowing the gender funding gap in other economies (Martiarena, 2019).

Gender biases in venture capital decisions contribute significantly to the funding disparity (Balachandra, 2020). Implementing standardized, gender-neutral criteria in evaluating startups can improve fairness and increase women's access to funding.

By implementing these strategies, Bangladesh can create a more inclusive investment ecosystem that fosters innovation and economic growth through gender-diverse entrepreneurship.

1.5.1 Promote Digital Literacy

Expanding digital literacy among women, especially in rural areas, is crucial for their participation in the tech startup ecosystem. Digital skills empower women economically, allowing them to start businesses, access global markets, and engage in remote work (Kumar, Kumar, & Devi, 2024).

The Bangladeshi government and NGOs should collaborate to expand digital literacy training for women in rural areas. Programs such as the Women ICT Freelancers Entrepreneurs Development Program have already demonstrated success in providing digital skills to women (Hamid et al., 2020).

Mobile learning platforms can be leveraged to deliver digital skills training to rural women. Research suggests that smartphones are increasingly becoming tools for informal learning, allowing women to develop digital skills in flexible, self-paced ways (Gautam & Rai, 2025).

Companies such as Microsoft and Google can collaborate with local organizations to create digital literacy workshops for women. International examples show that corporate-led digital education initiatives significantly improve women's ability to participate in tech-driven economies (Fauzia & Hidayati, 2023).

Financial literacy is a critical aspect of digital literacy. Studies show that training women in digital finance tools significantly improves their entrepreneurial success by giving them access to mobile banking and online financial services (Prabadevi et al., 2024).

By expanding digital literacy programs, Bangladesh can ensure that women, particularly in rural areas, have the tools to actively participate in the digital economy, fostering a more inclusive and dynamic startup ecosystem.

1.5.2 Monitor and Evaluate

Effective monitoring and evaluation (M&E) mechanisms are essential for measuring the impact of policies supporting women entrepreneurs. Research suggests that structured evaluation frameworks help identify gaps, measure progress, and refine strategies for maximum impact (Hassan & Njoroge, 2021).

Governments and organizations should establish clear, quantifiable metrics to track women's participation in entrepreneurship. This includes measuring business success rates, funding access, and employment generation (Wannamakok & Chang, 2020).

Periodic reviews of policies supporting women entrepreneurs should be conducted to ensure they remain effective and relevant to changing economic conditions (Bahmani-Oskooee, Galindo, & Méndez, 2012).

Effective M&E frameworks require collaboration between government agencies, financial institutions, and women entrepreneurs. Partnerships enhance data collection and improve decision-making (Orser, 2022).

Ensuring transparency through regular reporting on policy outcomes can build trust and promote accountability in women's entrepreneurship initiatives (Halabisky, 2018).

By integrating robust monitoring and evaluation frameworks, Bangladesh can ensure that policy interventions effectively support and sustain women entrepreneurs.

1.6 CONCLUSION AND STRATEGIC RECOMMENDATIONS

Empowering women in Bangladesh's tech startup ecosystem is a critical step toward achieving broader national goals of innovation, economic growth, and inclusive development. Although significant strides have been made, persistent barriers necessitate deliberate, sustained, and multifaceted strategies. Based on the evidence reviewed, this section provides strategic recommendations to enhance women's participation and leadership in Bangladesh's burgeoning tech startup space.

1.6.1 Expand Access to Digital and Entrepreneurial Education

One of the most fundamental steps in enhancing women's participation is expanding access to quality education and digital skills training. Early engagement in STEM fields is essential to create a pipeline of women entrepreneurs and leaders in technology (Faroque et al., 2024). Programs like the Women ICT Freelancers Entrepreneurs Development Program have demonstrated the efficacy of targeted skills training (Jaago Foundation, n.d.).

Thus, it is recommended that public-private partnerships be strengthened to deliver inclusive, gender-sensitive digital literacy programs. Mobile-based learning platforms, community tech hubs, and scholarships for women in tech education should be expanded, especially in rural and underserved areas (Gautam & Rai, 2025).

1.6.2 Address Financial Barriers through Gender-Lens Investment

A major barrier for women entrepreneurs remains unequal access to financing. Women-led startups consistently secure less venture capital compared to male-led counterparts (Davis, 2023). Bangladesh should prioritize the development of gender-lens investing initiatives, such as dedicated venture capital funds for women entrepreneurs and incentives for investors who back women-led startups (Debnath et al., 2020).

Organizations like BWIN provide a foundation, but greater scale and policy support are needed. Gender-neutral evaluation criteria must also be integrated into venture capital processes to combat unconscious bias in funding decisions (Balachandra, 2020).

1.6.3 Build Stronger Networks and Mentorship Platforms

Mentorship, peer networks, and role models play a crucial role in women's entrepreneurial success. The success of global initiatives like the Women's Startup Lab demonstrates that structured mentoring, international exposure, and peer learning can significantly enhance women's confidence and entrepreneurial capabilities (Women's Startup Lab, n.d.).

Bangladesh should expand similar mentorship networks domestically, linking experienced entrepreneurs and corporate leaders with emerging women founders. Annual bootcamps, pitch competitions, and international exchange programs specifically for women in tech can boost confidence and provide critical networking opportunities.

1.6.4 Strengthen Institutional and Policy Support

Although policy initiatives like Startup Bangladesh and ShePower are commendable, the execution and monitoring of such programs require greater rigor. A gender-responsive policy framework that integrates measurable goals and transparent reporting mechanisms is essential (Hassan & Njoroge, 2021).

The government should also streamline bureaucratic processes for women entrepreneurs seeking licenses, grants, and loans. Research indicates that simplifying access to financial resources can significantly boost women's entrepreneurial success (Bari et al., 2023).

1.6.5 Foster Inclusive Work Environments

Women entrepreneurs and tech professionals often face cultural barriers and subtle biases in the workplace. Developing inclusive organizational cultures that actively promote gender diversity is vital. Companies should adopt policies that encourage flexible working arrangements, mentorship opportunities for women, and gender-sensitization training for all employees.

Recognition programs like the Inspiring Women in Tech Awards should be expanded to showcase successful women leaders, thereby normalizing women's leadership in tech and entrepreneurship.

1.6.6 Promote Women's Leadership and Representation

Increasing women's leadership in startups and entrepreneurial support organizations is critical to driving systemic change. Research highlights that organizations with diverse leadership teams achieve higher innovation revenues and better financial outcomes (Lorenzo et al., 2018).

Quotas or targets for women's representation on startup boards, investment committees, and accelerator programs should be considered. Additionally, leadership development programs tailored for women can cultivate the next generation of tech startup leaders.

1.6.7 Leverage Technology for Social Impact

Women entrepreneurs in Bangladesh have already demonstrated how technology can address critical social challenges, such as access to healthcare, education, and financial services (Faroque et al., 2024). Supporting women-led tech ventures with a social impact focus should be a strategic priority.

Impact investment funds and social entrepreneurship incubators should specifically target women entrepreneurs solving pressing societal issues, thereby amplifying their influence in national development.

1.6.8 Ensure Robust Monitoring and Evaluation

Effective policies require robust monitoring and evaluation (M&E) mechanisms. The government and supporting organizations should establish clear indicators to track progress in women's entrepreneurship, such as startup survival rates, investment levels, job creation, and leadership representation (Orser, 2022).

Regular independent evaluations should be conducted to refine and adapt programs, ensuring they remain aligned with evolving challenges and opportunities.

1.6.9 Tackle Societal and Cultural Barriers

Beyond economic interventions, broader societal change is necessary to normalize women's participation in entrepreneurship. Campaigns promoting positive narratives of women entrepreneurs, public discussions on gender roles, and community engagement can shift societal perceptions (Solotaroff et al., 2018).

School curricula should also integrate modules on gender equality and entrepreneurship from an early age to foster a culture of inclusivity and innovation.

1.6.10 Foster Collaboration between Stakeholders

Finally, a coordinated effort by government agencies, private-sector actors, NGOs, academic institutions, and international organizations is crucial. Multi-stakeholder partnerships can pool resources, share knowledge, and create a holistic support system for women entrepreneurs.

Establishing a National Council for Women in Technology and Entrepreneurship, comprising representatives from various sectors, could help coordinate efforts, advocate for policy reforms, and promote best practices across the ecosystem.

1.6.11 Final Thoughts

Empowering women in Bangladesh's tech startup revolution is not just a moral imperative; it is an economic necessity. A tech ecosystem that embraces gender diversity will be more innovative, resilient, and globally competitive. By investing in education, financing, mentorship, policy reform, and cultural change, Bangladesh can unlock the vast potential of its women entrepreneurs, driving inclusive growth and setting a powerful example for the region.

Bangladesh stands at the cusp of a digital transformation. Ensuring that women are at the forefront of this change will be essential for building a future that is dynamic, equitable, and prosperous for all.

REFERENCES

The Asian and Pacific Training Centre for Information and Communication Technology for Development (APCICT). (n.d.). *WIFI*. The Asian and Pacific Training Centre for Information and Communication Technology for Development (APCICT). www.unapcict.org/ flags hip- pro gram mes/ wifi?form= MG0 AV3

Bahmani-Oskooee, M., Galindo, M. Á., & Méndez, M. T. (2012). Women's entrepreneurship and economic policies. In M. Á. Galindo & D. Ribeiro (Eds.), *Women's entrepreneurship and economics* (International Studies in Entrepreneurship, Vol. 1000, pp. 23–33). Springer. https://doi.org/10.1007/978-1-4614-1293-9_3

Balachandra, L. (2020). How gender biases drive venture capital decision-making: Exploring the gender funding gap. *Gender in Management: An International Journal, 35*(3), 261–273.

Bangladesh Women in Technology (BWIT). (n.d.). *About us*. Bangladesh Women in Technology (BWIT). www.bwit.org.bd/aboutus

Bangladesh Women in Technology (BWIT). (n.d.). *Projects | BWIT*. Bangladesh Women in Technology (BWIT). www.bwit.org.bd/projects

Bari, S., Ador, R. H., Akter, S. S., & Rahman, M. M. (2023). *The barriers of women entrepreneurs in accessing marketplace and financial supports: A study at Rajshahi City in Bangladesh*. Khulna University Studies.

Biniyog Bridhhi. (n.d.). *Scaling impact enterprises of Bangladesh (SIE-B)*. Biniyog Briddhi. www.sie-b.org/

Davis, D.-M. (2023, January 18). *Women-founded startups raised 1.9% of all VC funds in 2022, a drop from 2021*. TechCrunch. https://techcrunch.com/2023/01/18/women-founded-startups-raised-1-9-of-all-vc-funds-in-2022-a-drop-from-2021/

Debnath, G., Chowdhury, S., Khan, S., Farahdina, T., & Chowdhury, T. (2020). Role of women entrepreneurship on achieving sustainable development goals (SDGs) in Bangladesh. *The Business and Management Review, 10*(5).

Dhaka Tribune. (2022, March 28). *Bangladesh Angels Network launches BWIN to support women entrepreneurs, investors*. Dhaka Tribune. www.dhakatribune.com/business/266649/bangladesh-angels-network-launches-bwin-to

Faroque, A. R., Hossain, F. I., Mursalin, S., Jahangir, Z., & Islam, Q. T. (2024). *Empowering women in Bangladesh's tech startup revolution: A framework for participation, innovation, and growth.*

Fauzia, A., & Hidayati, S. (2023, November). Improving digital literacy of rural women entrepreneurs in Indonesia. In *Proceedings of the 11th International Conference on

Cyber and IT Service Management (CITSM 2023)*. IEEE. https://doi.org/10.1109/CITSM60085.2023.10455473.

The Financial Express. (2018, February 24). Government to initiate project for creating women ICT entrepreneurs. *The Financial Express.* https://thefinancialexpress.com.bd/economy/bangladesh/government-to-initiate-project-for-creating-women-ict-entrepreneurs-1519470345

Gautam, S., & Rai, I. M. (2025). Smartphones in the hands of rural women: Embracing informal learning for digital literacy. *Adult Learning, 36*(2), 120–130. https://doi.org/10.1177/10451595241263142

Google Women Techmakers. (n.d.). *We're building a world where all women can thrive in tech.* Google Women Techmakers. https://developers.google.com/womentechmakers

Guzmán, J., & Kacperczyk, A. (2018). Gender gap in entrepreneurship. *SRPN: Labor Performance Issues (Topic).*

Halabisky, D. (2018). Policy brief on women's entrepreneurship. *OECD SME and Entrepreneurship Papers.*

Hamid, R., Zukarnain, N., et al. (2020). Digital literacy among women entrepreneurs in rural areas. *Journal of* .of Contemporary Social Science Research, *4*(1), 87–93.

Hassan, M., & Njoroge, N. (2021). Influence of monitoring and evaluation planning on the performance of government programs: A case of Uganda Women Entrepreneurship Program in Busoga Sub-Region, Uganda. *International Journal of Scientific and Research Publications (IJSRP) 11*(10), 122–130. doi.org/10.29322/IJSRP.11.10.2021.p11816

Islam, B. (2024, March 9). Women's economic imperative: What do we need to do? *LinkedIn Pulse.* www.linkedin.com/pulse/womens-economic-imperative-what-do-we-need-bijon-islam-ytqtc/

Jaago Foundation. (n.d.). *Education and digital learning.* Jaago Foundation. https://jaago.com.bd/education-program

Kumar, R., Kumar, S., & Devi, M. (2024). Digital literacy: A pathway toward empowering rural women. *Global Journal of Women's Studies, 74.*

Lauprete, S. (2023, June 13). *Women are under-represented in the European startup ecosystem.* Boston Consulting Group. www.bcg.com/press/13june2023-women-are-under-represented-in-the-european-startup-ecosystem

Lorenzo, R., Voigt, N., Tsusaka, M., Krentz, M., & Abouzahr, K. (2018, January 23). *How diverse leadership teams boost innovation.* Boston Consulting Group. www.bcg.com/publications/2018/how-diverse-leadership-teams-boost-innovation

Martiarena, A., & Hart, M. (2019, January 8). State of the art review: Access to venture capital amongst female-led firms. (State of the Art Review No. 16). Enterprise Research Centre..

Mathew, A. (2021, December 16). *Code without barriers: Addressing the gender gap in tech.* Microsoft. https://news.microsoft.com/apac/2021/12/16/code-without-barriers-addressing-the-gender-gap-in-tech/

Orser, B. (2022). Building back better through feminist entrepreneurship policy. *International Journal of Gender and Entrepreneurship, 14*(4), 468–488. https://doi.org/10.1108/IJGE-05-2022-0089.

Prabadevi, M. N., Mary Auxilia, P. A., et al. (2024). Strategies for leveraging digital libraries to improve financial literacy among rural entrepreneurial women. *Indian Journal of Information Sources and Services, 14*(2), 28–33. https://doi.org/10.51983/ijiss-2024.14.2.05

Robb, A., Coleman, S., & Stangler, D. (2014). Sources of economic hope: Women's entrepreneurship. *Ewing Marion Kauffman Foundation Research Paper Series.*

Seshasayee, R. M. (2023, September 11). *An interview: Tina Jabeen and the fast-evolving Bangladesh entrepreneurship scene*. Cornell SC Johnson College of Business. https://business.cornell.edu/hub/2023/09/11/tina-jabeen-fast-evolving-bangladesh-entrepreneurship-scene/

Solal, I. (2019). The gender of money: Investor gender effects on early-stage venture financing. *ERPN: Labor Economics (Topic)*.

Solotaroff, J. L., Kotikula, A., Lonnberg, T., Ali, S., Pande, R., & Jahan, F. (2018). *Voices to choices: Bangladesh's journey in women's economic empowerment*. World Bank Group.

Startup Bangladesh Limited. (n.d.). *Who we are*. Startup Bangladesh Limited.

Stefanuto, L. (2023, March 27). *Only 15% of tech startup founders are female*. Startup Genome. https://startupgenome.com/articles/only-15-percent-of-tech-startup-founders-are-female

Wannamakok, W., & Chang, Y. (2020). Understanding nascent women entrepreneurs: An exploratory investigation into their entrepreneurial intentions. *Gender in Management: An International Journal, 35*(6), 553–566. https://doi.org/10.1108/GM-12-2019-0250.

Women's Startup Lab. (n.d.). *Home – Women's Startup Lab*. Women's Startup Lab. www.womenstartuplab.com/

2 Entrepreneurial Empowerment of Women through Digital Inclusion

The Impact of Mobile and Internet Penetration in Bangladesh

Farzana Nahid, Rakshanda Samin, and Saslina Kamaruddin

2.1 INTRODUCTION

The advent of digital technologies, mobile phones, and the internet has transformed business practices globally, acting as a catalyst for massive shifts in how businesses operate (Yadav & Pavlou, 2014). Mobile phones enable real-time communication and access to market information, making all aspects of businesses easier to conduct, from customer and supplier communication to cutting costs, streamlining processes, and adapting swiftly to dynamic market conditions (Berisha-Shaqiri, 2015; Tharu & Yadav, 2018). Similarly, the internet and online platforms acting as a single database and connection point has improved efficiency, expanded market reach, and facilitated online transactions. From marketing, operations, financial management, and more, digitization has revolutionized business processes and fostered entrepreneurs who would experience the benefits of the brand-new paradigm – especially so for marginalized groups with previously limited access to resources, such as women (Maier & Nair-Reichert, 2008).

Women across the globe have leveraged the use of digital technologies to overcome traditional barriers and achieve financial independence (Alam & Zuma, 2023). Bangladesh, having experienced impressive economic growth over the past few decades, has also seen a significant increase in women's participation within the entrepreneurial scene (Uddin & Barua, 2024). In Bangladesh, like many other conservative societies, patriarchal norms have traditionally confined women to domesticated roles, limiting their participation in economic activities beyond their homes (Tabassum, 2018). Gender disparities in education and resources have further constrained women's opportunities to establish themselves as economically independent and have

DOI: 10.1201/9781003544838-2

greater agency (Kabeer, 2017). However, as mobile phones and the use of the internet became widespread, availability of digital technology and internet access leveled the playing field for many women by offering them avenues for growth.

Research shows that as online resources and platforms educate and better equip women to regain autonomy in modern society, they become able to start their own businesses with minimal investment, bypassing barriers to entry within the world's male-dominant society (Jeevitha et al., 2023). In Bangladesh, where small- and medium-sized enterprises (SMEs) are popularly considered to be the backbone of the economy, digital technologies have given women entrepreneurs the power to make operations more efficient, connect with new clients, and expand the range of products they offer, making women-owned SMEs a noteworthy section of businesses using e-commerce platforms and online social media websites to reach their audience (Ahmed et al., 2023).

Digital technologies have therefore crafted a pathway for women in oppressed societies to achieve independence and reclaim rights by attaining financial autonomy and ensuring decision-making power. Besides gaining access to countless online resources to improve digital and financial literacy and conduct business processes, it has allowed them to connect with similar communities and seek encouragement, inspiring the growth of women-owned SMEs within the country (Tabassum, 2018). However, there is still a need to understand to what extent these technologies are able to aid women entrepreneurs amidst longstanding cultural norms and infrastructure challenges, while also addressing existing constraints within the world's primarily male-dominated society (Afrah & Fabiha, 2017). This chapter has therefore attempted to look deeper into that gap and examine how mobile phones and internet access have successfully empowered women entrepreneurs in Bangladesh. By examining their unique experiences and the obstacles they have faced, this research aims to enhance our understanding of digital inclusion and gender disparity in an entrepreneurial context and contribute to existing literature on the subject, shedding light on how digitization can promote gender equality and economic progress.

2.2 LITERATURE REVIEW

To say that the internet has turned the business landscape on its head over the past few decades would be an understatement. From business operations to marketing strategies, the adoption of the internet and technology has revolutionized both business-to-business (B2B) and business-to-consumer (B2C) paradigms by first enabling better information exchange and customer support, then reducing costs and streamlining processes, improving connections with suppliers and customers, and ultimately increasing overall business efficiency (Berisha-Shaqiri, 2015; Tharu & Yadav, 2018; Castells, 2002). If we consider the broader implications, internet and digital technology have played a huge part in driving social and economic development as well (Singh et al., 2025; Tharu & Yadav, 2018). Castells (2002) discussed how digitizing the global business landscape has inspired new business models and increased value creation by shifting from traditional processes to more dynamic, technology-driven ones. When we narrow the scope down towards developing economies, we find studies like that of Islam and Nasira (2017) which mention how the internet has notably opened up newer avenues for growth and expansion of SMEs in developing countries. Incorporating information and communication technology (ICT) within the SMEs ecosystem in Bangladesh has led to increased productivity and growth (Islam & Nasira, 2017). According to Dutta et al. (2020), Bangladesh also follows global trends when embracing significant shifts towards digital marketing and e-commerce. Furthermore, digital inclusion initiatives have improved access to education, healthcare, and financial services, contributing to overall socio-economic development (Berisha-Shaqiri, 2015; Rahayu & Day, 2017).

The digital divide, however, still remains an issue, especially for marginalized groups such as women. Studies have shown that women are less likely to have access to the internet and digital technologies compared to men, which adds to existing inequalities (Hafkin & Huyer, 2007; Berisha-Shaqiri, 2015). Studies discuss that despite the availability of technological devices, women still face barriers in both job searching and business expansion due to pre-established male dominance and intense competition (Abraham, 2009; Burrell, 2010; Voufo et al., 2025). This digital gender gap limits women's participation in the digital economy and hinders their potential for economic empowerment and needs to be addressed through initiatives that target both digital literacy and infrastructure challenges (Tharu & Yadav, 2018; Singh et al., 2025). On the other hand, in spite of all these divides, the role the internet and technology have played in helping women prosper against all odds is undeniable.

2.3 WOMEN, TECHNOLOGY, AND DIGITIZATION: THE INTERNET'S ROLE IN EMPOWERING WOMEN

With the vast majority of the world's societies being patriarchal, women around the world have been faced with various setbacks and challenges since the beginning of time. These setbacks manifest in many forms ranging from limited educational access to restricted financial autonomy and lack of decision-making power. Cultural norms are often known to restrict women's roles to homemaking, giving opportunities for education and economic participation far less importance than domestic roles (Kabeer, 2017). In South Asian countries, societal and cultural practices place heavy restrictions on women that keep them almost strictly within domestic roles, limiting them from pursuing careers or entrepreneurial ventures (Nussbaum, 2000; Corrêa et al., 2024). It is also important to note that beyond measurable economic achievements, women empowerment must also entail increasing opportunities available to them and promoting social transformation by allowing women to better understand the political frameworks that hold them back and to have more agency over their lives (Mayoux, 2001; Cornwall & Edwards, 2014).

In light of such conservative social frameworks and setbacks, internet and digital technologies have opened up avenues for these home-bound or confined women.

Digital literacy then ends up playing a crucial role in empowering women by allowing access to information, learning, and skills training that were previously unavailable. The internet and social media become platforms for education and development, helping women to bridge the knowledge gap and participate more fully in the economy (Alam & Zuma, 2023; Sowmya & Pai, 2025). A study by Maier and Nair-Reichert (2008) brings to light how e-commerce platforms have empowered women by providing them with the tools and knowledge necessary to start and manage their very own businesses. Online resources and communities offer support and mentorship, helping women to build confidence and competence from within their homes. Additionally, digital technologies allow access to global audiences, letting women expand their businesses beyond local limitations (Yadav et al., 2022; Afrah & Fabiha, 2017). Overa (2006) discusses that a deep dive into the impact mobile phones have had on women empowerment will reveal a focus on economic benefits, improved time management and coordination, and strengthened social identity. In Bangladesh, initiatives like "Digital Bangladesh" have focused on increasing digital literacy among women, upholding its importance in promoting gender equality and economic participation (Akther et al., 2025).

Through the internet, women are able to access microfinance and crowdfunding platforms, which solve the primary funding problem that startups face (Alam & Zuma, 2023; Afrah & Fabiha, 2017). Many of these platforms often cater specifically to women entrepreneurs and their unique challenges and go on to offer tailored solutions (Yadav et al., 2022; Sowmya & Pai, 2025). Besides social media, e-commerce, and more specifically f-commerce (Facebook-based e-commerce), platforms have played an unparalleled role in creating opportunities for women-led businesses by providing them with nearly cost-free opportunities to sell products and services online, reaching a broader customer base and increasing their income potential (Tabassum, 2018; Jeevitha et al., 2023). Rahamn (2023) pointed out that the flexibility and low barriers of entry of e-commerce or f-commerce platforms allow women to easily juggle their household responsibilities and economic activity. Afrah and Fabiha (2017) noted that these online platforms remove the need for women in physical, male-dominant market spaces, essentially protecting them from the oppression and stigma. Santra et al. (2020) also echo that online platforms make it easier for women to balance their domestic and professional lives.

Vast literature outlining women's success stories can be found. Women-owned businesses on e-commerce platforms in India saw a 45% increase in sales during the COVID-19 pandemic (Santra et al., 2020). Additionally, statistics quoted by Yadav et al. (2022) show a staggering 60% growth in women-owned businesses in developing countries within the first year of operation. This status quo not only improves the economic status of women but also contributes to broader socio-economic shifts by challenging and reshaping patriarchal norms (Nussbaum, 2000; Alam & Zuma, 2023). Studies show that penetration of mobile and the internet has successfully reduced gender gaps in financial inclusion and economic participation. The most important point to note, however, might be that digital literacy and internet access have been pivotal in enhancing women's self-assuredness and agency, enabling them to challenge conservative beliefs and outdated practices (Maier & Nair-Reichert, 2008; Afrah & Fabiha, 2017).

2.4 DIGITIZATION, ENTREPRENEURSHIP, AND BANGLADESHI WOMEN: BREAKING TRADITIONAL BARRIERS

Though cultural and societal norms in Bangladesh have historically limited women's participation in economic activities, through digitization, the country's women have been able to break the glass ceiling. Women-led SMEs have benefitted vastly from online platforms that made tools to manage businesses more effectively and reach broader markets more accessible to women, and women entrepreneurs who embraced digital tools experienced a 30% growth in business efficiency and customer satisfaction (Akther et al., 2025). Mobile and internet have granted women financial autonomy through digital finance solutions like mobile banking and online payment systems, making access to credit and investment opportunities easier. Women entrepreneurs using mobile banking services experienced a 40% increase in business revenues (Nugroho & Chowdhury, 2015), and digital platforms have made crowdfunding and peer-to-peer lending (Tariq, 2025) a lot more accessible. Furthermore, mobile banking has significantly reduced the gender gap in financial inclusion, with 60% of women entrepreneurs accessing loans and credit through digital platforms (Oyasor, 2026; Nugroho & Chowdhury, 2015).

Online platforms offer anonymity and flexibility, letting women conduct business activities from their homes, which is crucial in a society where women's mobility and social interactions are highly restricted. Digital entrepreneurship allowed women to bypass societal restrictions, with 80% of survey respondents in a study indicating that online business enabled them to work from home and manage family responsibilities effectively (Al-Amin et al., 2024; Maier & Nair-Reichert, 2008). The strides made by Bangladeshi women who employ mobile and internet to do business are undoubtedly noteworthy. For example, women in rural areas have used mobile technologies to sell agricultural products and handicrafts online, balancing domestic responsibilities with entrepreneurial activities (Tabassum, 2018; Maier & Nair-Reichert, 2008). In the garments industry, essentially the most important industry in Bangladesh, women-led businesses that adopted digital marketing strategies saw a 50% increase in exports during the pandemic (Islam et al., 2022; Akther et al., 2025). A study on the use of internet in women-led SMEs in Bangladesh during the pandemic proved how they were able to manage businesses fully online, with 70% of women-led SMEs maintaining or increasing their customer base through digital marketing. The Bangladesh Women Chamber of Commerce and Industry (BWCCI) found in a study that 65% of women entrepreneurs who adopted digital platforms reported an increase in sales during the COVID-19 pandemic (Uddin & Barua, 2024; Islam & Nasira, 2017).

The significance of mobile and internet penetration on women's entrepreneurial empowerment in Bangladesh therefore cannot be emphasized enough. It has made access to business-related tools, financial services, supportive communities, and global markets much easier, enabling women to overcome traditional barriers and achieve economic success. As mentioned by Showkat et al. (2025), access to digital finance solutions has enabled women in Bangladesh to initiate and grow their enterprises, leading to economic stability and empowerment. According to Begum (2024), women in Bangladesh who engage in digital entrepreneurship have reported a 30% increase in household income and greater decision-making power within their families. The success stories of women-led SMEs and tech startups illustrate the transformative potential of digitalization, promoting gender equality and economic development (Akther et al., 2025; Oyasor, 2026).

2.5 METHODOLOGY

This chapter used a qualitative research approach to collect data in order to explore the experiences of women entrepreneurs in Bangladesh in terms of using mobile phone and internet. Through in-depth interviews with approximately 30 SME owners from diverse socio-economic backgrounds and from different districts of Bangladesh such as Dhaka Chittagong, Sylhet, Cox's Bazar, and Rajshahi, this research aims to capture insights into how mobile phones and the internet have influenced their entrepreneurial journeys. Through the semi-structured interviews with women entrepreneurs from diverse sectors and geographical locations across Bangladesh, this chapter captures detailed insights into how mobile phones and the internet have empowered the women, especially by influencing their business strategies, market access, decision-making autonomy, and overall economic and societal independence.

By using a qualitative approach while drawing on feminist methodology (Ramazanoglu & Holland, 2002), this study prioritizes the voices and experiences of women entrepreneurs, ensuring their perspectives shape policy recommendations and development strategies. By using triangulation method (as interviews, literatures, observation), this study is not only providing the beneficial scenarios created through mobile and internet for the women entrepreneurs of Bangladesh, but also capturing actionable recommendations in order to enhance digital literacy programs and to expand internet infrastructure so these women can mitigate the challenges they are facing now and can further utilize the digital platforms for their overall development.

2.6 FINDINGS

From the interviews it is very much evident that smart mobile phones with internet have given the women of Bangladesh the opportunity to easily start a business while being at home, taking care of households, and so on. In fact, the data clearly revealed how mobile phones and internet connectivity have transformed their overall lives by helping the women to be financially independent and skilled, facilitating networking and collaboration, fostering social empowerment, improving access to market and operational efficiency in business, and so on. However, some challenges also got revealed, such as digital divide, cybersecurity threats, and socio-cultural norms which continue to pose barriers to women's full participation in the digital economy.

The interviews reveal that most of the women entrepreneurs frequently use a variety of digital tools to streamline their business operations. For designing visually pleasing social media posts, Canva is a popular choice due to its user-friendly interface and large template library, whereas Capcut is mostly used for video editing and reels. They also use Google Sheets as it serves as a versatile tool for managing various aspects of their business, including order tracking, inventory management, and daily journaling. For presentations and data analysis, they mostly use

MS PowerPoint and MS Excel. To drive product promotion, they rely on Facebook Ads and Google AdWords. Meanwhile, Google Drive and Dropbox are preferred for storage, sharing, and ease of access provided by these platforms. Nevertheless, in terms of communication and collaboration, they commonly use platforms such as Facebook, WhatsApp, Messenger, Viber, Google Meet, and Gmail. These tools facilitate seamless interactions with clients, team members, and partners, enabling efficient management of their business activities.

From the interview, women entrepreneurs did express their happiness to get easy market access and fast business expansion by using internet-based platforms for business. For example, Ms. Tabassum, an online jewelry owner, said, "Due to digital technology I can manage my business activities alone, with no distance or time constraints. My business has unlimited boundaries through internet." Ms. Zafrin, who sells agricultural products, said, "Digital platforms have significantly contributed in the growth of my business. I started with almost nothing, but now I am able to promote and ship my products nationwide and internationally." An herbal product maker, Ms. Jahangir, mentioned, "Internet is a blessing. Facebook is the main buying platform in Bangladesh where people choose to buy based on reviews and vlogs." A jute product maker, Ms. Shoili, said, "Digital platforms have enhanced my business growth and sustainability. Unlike offline methods, online platform especially social media allowed me to reach more audience anywhere in the world, with 24/7 availability and cost-effective advertisement facility."

The majority of the women emphasized the ***operational efficiency*** gained from mobile phones and internet-based platforms. An online interior designer, Ms. Sultana, stated, "I use mobile phone to take orders, to communice with customers and suppliers, source products, and to save money by doing business without having a physical shop." Ms. Jannat, a fashion house owner, noted, "Mobile phone and internet simplify business operations. I share product pictures and videos, and can do video calls for supply chain communication." Ms. Ishrat, a boutique owner, mentioned, "Since Covid, life is all about mobile phone and internet. Although I have a brick-and-mortar store, my product promotion is done via Facebook, which is cost-effective and reaches all my customers." Ms. Safa, an herbal product maker, said, "Earlier, the only way to do business was through a brick-and-mortar store. With a smartphone, I save money on promotion, communication, and operations, while online shopping and tracking boost customer satisfaction." An online boutique owner, Ms. Sajila, stated "Virtual meetings through Google Meet or Zoom help me overcome distance and time barriers in my business communication." Ms. Afia, an interior designer, highlighted, "Online platforms and apps like QuickBooks and Zoom have replaced manual accounting and face-to-face meetings, and thus making business management more efficient." Ms. Zinia Afrin, an agricultural product seller, observed, "Using the phone and Facebook for orders and payments is easy and cost-efficient. I can also share product growth and packaging videos, enhancing customer trust."

Financial independence is another major realization among these women. Many women highlighted how mobile phones and internet platforms have enabled them to achieve financial independence while managing other responsibilities: Ms. Tanzu, an e-commerce owner, said, "With just a mobile phone and internet, I could start a business, solve my financial problems, build self-identity, and contribute to my family's earnings." Ms. Banu, a crafts owner, noted, "Facebook allows customers to find and order my products easily. Mobile banking services like Bkash or Nagad simplify transactions, letting me manage my business while taking care of my family."

Women entrepreneurs have found internet-based platforms like Facebook, LinkedIn, and WhatsApp groups significantly useful in order to ***enhance their networking and collaboration opportunities.*** The majority of the women expressed similar thoughts about it. They mentioned that the internet has been a blessing for them in expanding their business network. Joining various local and international entrepreneurs' groups and associations on platforms like Facebook and WhatsApp has expanded their business boundaries. Not only that, Facebook allowed them to participate in online courses, workshops, and trainings.

The women also expressed their satisfaction in ***achieving decision-making autonomy*** by using mobile phones and internet while highlighting benefits in time, cost, and energy savings. An online boutique owner, Ms. Shahana, stated,

> Earlier I had to go everywhere and struggle with time management. But now I can manage household and business tasks parallelly by using mobile phone. In fact, I can order raw materials, take customer orders, deliver products, handle groceries and bill payments online, which save both my time and energy.

A bamboo and cane product maker, Ms. Kahdija, said,

> Digital inclusion through the use of mobile phones and internet has enabled me to make independent business decisions. I can easily learn about market trends, competition, and customer needs, can get instant access to economy, currency exchange rates, and other important economic information, can communicate directly with customers in Social and can collect feedback from them through surveys. I could easily create an online store and reach customers worldwide, could transact quickly and securely using mobile banking and online payment systems, can manage tasks with tools like Asana or Trello, communicate with my team via WhatsApp or Zoom, and reach targeted customers using digital marketing tools.

Nevertheless, the women entrepreneurs have expressed significant feelings of ***social empowerment*** through the use of mobile and internet platforms: as they mentioned, "I belong to a middle-class family, and got married early. Mobile and internet have given me the access to many learning resources which allowed me to continuously learn business topics without going to school/institutions." An online food-catering service owner, Ms. Naznin, said, "I can't imagine my life without mobile and internet. For all women entrepreneurs in Bangladesh, it's a jackpot. Women of Bangladesh could gain finance independence and a better social identity and confidence while living in a male dominated society." A women entrepreneur from Sylhet stated, "Traditional barriers in Sylhet make it difficult for women to go out for work or run a business. Internet has helped me to overcome this, allowed me to do business online, sell my products without leaving the house." A food ingredients seller, Ms. Tori, noted,

> Working offline is traditional, labor intensive, and requires large capital. For Bangladeshi women, who are expected to prioritize household duties, mobile and internet platforms offer a path to financial independence, enabling us to do business online even as housewives or mothers.

All of them agreed with the fact that internet-based learning is easy, which helped them to ***expand their skillsets and capacity-building.*** Ms. Tori, a craft business owner, stated, "By using internet-based platforms I have done a lot of training from home, like calligraphy, photography, spoken language, many more things that were not easily possible for me offline. This is how I could develop my skills."

Well, the interviews also reveal several challenges the women entrepreneurs face by using internet, Facebook in particular. They mentioned challenges concerning cyber security, technology integration, data privacy, and reliance on a stable internet connection. Almost all of the women interviewed mentioned cyber security risks. They commonly mentioned that sensitive business information is at risk of being stolen or compromised through hacking attempts. Additionally, they mentioned phishing scams which target personal information via fraudulent emails or messages. To mitigate these risks, these women should know to invest into robust cyber security measures, which include using unique passwords and making sure to change them regularly. Furthermore, keeping software and operating systems up to date will provide protection against vulnerabilities that may be exploited by malicious parties.

Data privacy is another critical issue mentioned by them. Safeguarding customer information is crucial because it is both a legal obligation and maintains credibility and trust of the seller. Implementing encryption for sensitive data can be considered an essential layer of protection here. In addition, establishing a clear privacy policy can be helpful to ensure transparent and secure data handling practices.

Some women mentioned technology integration as its own set of challenges. Many living outside metropolitan cities don't know the right way of using mobile and internet technology. Besides, the little they know is not enough for them to conduct business online. They also need to have cyber security training to recognize and respond to potential threats besides having technology training to effectively utilize new tools and systems. Knowing about regular backups of data, stored in secure locations, can further safeguard their businesses against data loss due to technical issues or security breaches.

Lastly, the dependence on a stable internet connection is another concern of the women entrepreneurs. They did mention that having a reliable and high-speed internet connection is crucial for seamless business operations. In particular, women living in villages or small cities don't have seamless internet connection and electricity to charge the mobile properly. To minimize disruptions, continuous connectivity should be ensured.

By proactively addressing these challenges with the outlined strategies, women entrepreneurs' businesses can enhance their resilience against cyber threats, protect their valuable data, and smoothly integrate new technologies in order to maintain operational continuity.

2.7 ANALYSIS

In Bangladesh, women entrepreneurs have seen life-changing opportunities through the use of mobile phones and the internet, which have worked in their favor in various ways. Women have had access to these much-needed tools in their quest for financial freedom. With easy access to learning and greater self- as well as business exposure through the use of the internet, women have become more capable of managing personal and professional lives side by side. Economic empowerment has provided them with the opportunity to participate in the economic development of their families and the broader society and subsequently command more respect and social status within their communities. Moreover, e-commerce and other online businesses have also made it possible for these women to participate in activities that support women's employment generation and a more active economy. Hence, it is the use of the internet and mobile phones that has led to improving their livelihoods, making positive impacts in their corners, and enhancing economic advancement.

The observation from these interviews with other women in Bangladesh fills in the realization that digital inclusion is not only a way of advancing women's entrepreneurship in terms of women-owned SMEs, but also one of the tools for achieving certain socio-economic development objectives. Women are able to learn business, develop product and service ideas, adapt to the market, and grow the business by the use of mobile phones and internet since it reduces information asymmetry in the market and accelerates the exchange of ideas. In addition to this, mobile platforms also enhance women's empowerment by allowing them to break free from the stereotypes of women and promoting their participation in the economy.

The analysis further highlights that, in this case, empowerment through the use of the internet also promotes socio-economic development of women business owners in Bangladesh. It has been noted that mobile phones can assist women in breaking these constraints by providing greater mobility and access to the business network (Buvinic & Furst-Nichols, 2016). Additionally, as stated within the interviews, it is very apparent that the digital technologies, due to their capabilities, assist in the acquisition of skills, enhance market entry, and foster innovative ideas, all of which enhance possibilities of advancing economic development and gender equity (Hafkin & Huyer, 2007).

Furthermore, challenges like digital literacy gaps, cyber security risks, and inadequate infrastructure remain significant barriers to maximizing the benefits of digital inclusion for women entrepreneurs (Singh et al., 2025; Hafkin & Huyer, 2007). In fact, barriers such as the differences in social classes when it comes to smartphone and internet use, hate speech online, and online harassment, which includes threats to kill, remain serious obstacles to overcome. Addressing these challenges requires concerted efforts from policymakers, development practitioners, and private sector stakeholders to promote inclusive digital strategies and gender-responsive policies (Yadav et. al., 2022). True sustainable development therefore requires collaborative efforts from government, private sector, and also the civil society to create the kind of environment that helps women entrepreneurs thrive and contribute meaningfully to Bangladesh's economic resilience.

2.8 CONCLUSION

In Bangladesh, where cultural norms limit women's movement and decision-making, digital technology can offer a way for them to gain more control over their economic lives. But it's still not clear exactly how these technologies are helping women entrepreneurs, especially in a context like Bangladesh where social expectations and weak infrastructure can still get in the way.

This chapter looks into how women entrepreneurs are using mobile phones and the internet to break through those barriers. By sharing their experiences, it gives a clearer sense of what opportunities they're finding and what challenges still hold them back. It adds to the broader conversation about how digital inclusion can empower women.

It also argues that the approach to digital inclusion in Bangladesh needs to be more comprehensive, focusing on access to mobile phones, internet, and building digital skills. If we prioritize giving women more access to online resources, Bangladesh could build a more inclusive economy where women entrepreneurs play a bigger role in driving growth. Over time, narrowing the digital gap could lead to real progress in gender equality and help women take on stronger roles in shaping the country's future.

2.9 STUDY CONTRIBUTION

This chapter underscores the necessity of integrating digital strategies into gender empowerment frameworks in order to realize the full potential of women entrepreneurs in Bangladesh. Policymakers and development practitioners can accelerate progress towards gender equality and sustainable development goals by encouraging inclusive and equitable access to digital tools and promoting business ecosystems that are more supportive and understanding towards women. Investments in digital infrastructure, education, and entrepreneurship support programs are also important in order to harness the transformative power of technology for women's economic empowerment.

REFERENCES

Abraham, K. B. (2009). The names in your address book: Are mobile phone networks effective in advocating for women's rights in Zambia? In I. Buskens & A. Webb (Eds.), *African women and ICTs: Investigating technology, gender and empowerment*. Zed Books/IDRC.

Afrah, S. H., & Fabiha, S. T. (2017). Empowering women entrepreneurs through information and communication technology (ICT): A case study of Bangladesh. *Management*, *7*(1), 1–6.

Ahmed, N., Chowdhury, A. M., Urmi, T., & Jamal, L. (2023). Impact of socio-economic factors on female students' enrollments in science, technology, engineering and mathematics and workplace challenges in Bangladesh. *American Behavioral Scientist*, *67*(9), 1104–1121. https://doi.org/10.1177/00027642221078517

Akther, W., Arif, A. A., & Islam, M. S. (2025). Women entrepreneurship development in the digital age: Challenges and opportunities in Bangladesh. *International Journal of Managerial Studies and Research*, *13*(6), 47–56.

Al-Amin, S., Rony, A. K., & Rahman, S. S. (2024). Impact of women entrepreneurship on digital entrepreneurship: A focus on women's participation in Bangladesh. *Journal of Jilin University (Engineering and Technology Edition)*, *43*(10), 137–150. https://doi.org/10.5281/zenodo.13969563

Alam, J., & Zuma, S. K. (2023). Internet and women: Does the internet influence women's empowerment. *Journal of International Business and Management*, *6*(4), 1–19. https://doi.org/10.37227/JIBM-2023-04-5812

Begum, R. (2024). *Women Eetrepreneurs In Bangladesh: Challenges and Opportunities* (Doctoral dissertation, © University of Dhaka). http://repository.library.du.ac.bd:8080/ xmlui/bitstream/handle/123456789/3033/Marketing%20PhD%20%282%29.pdf?sequence=1&isAllowed=y

Berisha-Shaqiri, A. (2015). Impact of information technology and internet in business. *Academic Journal of Business, Administration, Law and Social Sciences*, *1*(1), 73–79.

Burrell, J. (2010). Evaluating shared access: Social equality and the circulation of mobile phones in rural Uganda. *Journal of Computer-Mediated Communication*, *15*(2), 230–250. https://doi.org/10.1111/j.1083-6101.2010.01518.x

Buvinic, M., & Furst-Nichols, R. (2016). Promoting women's economic empowerment: What works? *The World Bank Research Observer*, *31*(1), 59–101. https://doi.org/10.1093/wbro/lku013

Castells, M. (2002). *The Internet galaxy: Reflections on the Internet, business, and society*. Oxford University Press. https://doi.org/10.1093/acprof:oso/9780199255771.001.0001

Cornwall, A., & Edwards, J. (2014). Introduction: Negotiating empowerment. In A. Cornwall & J. Edwards (Eds.), *Feminisms, empowerment and development: Changing women's lives*. Zed Books.

Corrêa, V. S., Lima, R. M. D., Brito, F. R. D. S., Machado, M. C., & Nassif, V. M. J. (2024). Female entrepreneurship in emerging and developing countries: A systematic review of practical and policy implications and suggestions for new studies. *Journal of Entrepreneurship in Emerging Economies*, *16*(2), 366–395. https://doi.org/10.1108/JEEE-04-2022-0115

Dutta, S., Lanvin, B., & Wunsch-Vincent, S. (2020). *The global innovation index 2020: Who will finance innovation?* Cornell University, INSEAD, and WIPO. www.wipo.int/edocs/pubdocs/en/wipo_pub_gii_2020.pdf

Hafkin, N., & Huyer, S. (2007). Women and gender in ICT statistics and indicators for development. *Information Technologies and International Development*, *4*(2), 25–41. https://doi.org/10.1162/itid.2008.00006

Islam, M. A., Abbott, P., Haque, S., Gooch, F., & Akhter, S. (2022). The impact of Covid-19 on women workers in the Bangladesh garment industry. *The University of Aberdeen and the Modern Slavery and Human Rights Policy and Evidence Centre (Modern Slavery PEC), UK, 12*(5), 226–39.

Islam, M., & Nasira, S. (2017). Role of technology on development of SME: Bangladesh perspective. *Journal of Entrepreneurship & Management*, *6*(1), 1–11.

Jeevitha, P., Priya, R. K., Rathna, G. A., & Tamilselvi, R. (2023). Challenges faced by women entrepreneurs (Conference paper). https://www.researchgate.net/publication/367296749_CHALLENGES_FACED_ BY_WOMEN_ENTREPRENEURS

Kabeer, N. (2017). Economic pathways to women's empowerment and active citizenship: What does the evidence from Bangladesh tell us? *Journal of Development Studies*, *53*(5), 649–663. https://doi.org/10.1080/00220388.2016.1205730

Maier, S., & Nair-Reichert, U. (2008). Empowering women through ICT-based business initiatives: An overview of best practices in e-commerce/e-retailing projects. *Information Technologies and International Development*, *4*(2), 43–60. https://doi.org/10.1162/itid.2008.00007

Mayoux, L. (2001). *Jobs, gender and small enterprises: Getting the policy environment right*. International Labour Organization.

Nugroho, L., & Chowdhury, S. L. K. (2015). Mobile banking for empowerment Muslim women entrepreneur: Evidence from Asia (Indonesia and Bangladesh). *Tazkia Islamic Finance and Business Review*, *9*(1), 83–100. https://doi.org/10.30993/tifbr.v9i1.79

Nussbaum, M. (2000). *Women and human development: The capabilities approach*. Cambridge University Press. https://doi.org/10.1017/CBO9780511841286

Overa, R. (2006). Networks, distance, and trust: Telecommunications development and changing trading practices in Ghana. *World Development*, *34*(7), 1301–1315. https://doi.org/10.1016/j.worlddev.2005.11.015

Oyasor, E. (2026). Impact of digital financial services on women's empowerment and financial inclusion in Asian countries. *Revista Científica Profundidad Construyendo Futuro, 24*(24), 35–48. https://doi.org/10.22463/24221783.5278

Rahamn, M. M. (2023). *Impact of social media marketing on consumer purchase action: A case study of SME consumers in Bangladesh* [Doctoral dissertation, University of Wales Trinity Saint David].

Rahayu, R., & Day, J. (2017). E-commerce adoption by SMEs in developing countries: Evidence from Indonesia. *Eurasian Business Review, 7*(1), 25–41. https://doi.org/10.1007/s40821-016-0044-6

Ramazanoglu, C., & Holland, J. (2002). *Feminist methodology: Challenges and choices.* SAGE Publications. https://www.proquest.com/scholarly-journals/feminist-methodology-challenges-choices/docview/197718144/se-2

Santra, S., Das, S. K., & Biswas, S. (2020). Impact of COVID-19 pandemic on women entrepreneurs: An Indian study. In *Globsyn Management Conference Proceedings.* Globsyn Business School, Kolkata, India

Showkat, M., Nagina, R., Baba, M. A., & Yahya, A. T. (2025). The impact of financial literacy on women's economic empowerment: Exploring the mediating role of digital financial services. *Cogent Economics & Finance, 13*(1), 2440444. https://doi.org/10.1080/23322039.2024.2440444

Singh, S., Rahul, K., Paliwal, M., Wani, I. A., & Suri, S. (2025). Gendering the digital divide: A systematic review of women's digital inclusion challenges and emerging research directions. *Digital Transformation and Society, 4*(4), 503–531. https://doi.org/10.1108/DTS-04-2025-0083

Sowmya, B., & Pai, R. (2025). Digital inclusion of women entrepreneurs in the unorganized sector: A systematic review. *Advances in Consumer Research, 2*, 71–85.

Tabassum, T. (2018). Facebook-based women entrepreneurship in Bangladesh: Opportunities and challenges. *Bangladesh Journal of Public Administration, 26*(2). https://doi.org/10.36609/bjpa.v26i2.22

Tariq, M. U. (2025). Empowering women entrepreneurs: Innovative crowdfunding and alternative financing models for sustainable business growth. In N. Katper (Ed.), *Crowdfunding and alternative financing models for women entrepreneurs* (pp. 1–34). IGI Global Scientific Publishing. https://doi.org/10.4018/979-8-3693-9193-8.ch001

Tharu, M., & Yadav, R. G. (2018). Effects of technological development through electronic communication in women empowerment. *International Journal of Humanities and Social Sciences, 7*(5), 93–100.

Uddin, N., & Barua, P. (2024). Women's economic empowerment and covid-19 pandemic: A study on women entrepreneurs in Bangladesh. *Journal of Economy and Technology, 2*, 258–268. https://doi.org/10.1016/j.ject.2024.08.006

Voufo, C. W., Gala, P. M., & Rodriguez, M. R. B. (2025). Analyzing the gender digital skills divide in Sub-Saharan Africa: Current state and contributing factors. *Education Working Papers*, (203094). https://thedocs.worldbank.org/en/doc/a607bb6e3b76d2be0f3db8db34dcf73e-0140022025/related/7EDU-WP-16-Analyzing-the-gender-digital-skills-divide-in-SSA.pdf

Yadav, H., Paliwal, M., & Chatradhi, N. (2022). Entrepreneurship development of rural women through digital inclusion: Examining the contributions of public programs. In *Inclusive businesses in developing economies: Converging people, profit, and corporate citizenship* (pp. 287–309). Springer. https://doi.org/10.1007/978-3-031-12217-0_14

Yadav, M. S., & Pavlou, P. A. (2014). Marketing in computer-mediated environments: Research synthesis and new directions. *Journal of Marketing, 78*(1), 20–40. https://doi.org/10.1509/jm.12.0020

APPENDIX

Semi-Structured Interview Questions:

1. **Can you describe how you make use of mobile phones and the internet in business operations?**
   - Probe: How has access to technology and mobiles facilitated communication with your suppliers, customers, and stakeholders?
   - Probe: In what ways has internet connectivity boosted your ability to reach new customer bases and allow business expansion?

2. **What are the challenges that you faced while integrating digital technologies into business practices?**
   - Probe: How have you addressed issues such as digital literacy or access to reliable internet services?
   - Probe: Have you faced any cybersecurity concerns, and how do you mitigate these risks?

3. **How has digital inclusion through mobile phones and internet access empowered you to make independent business decisions?**
   - Probe: Can you provide examples of how you use online resources to research market trends or negotiate business agreements?
   - Probe: In what ways has digital information accessibility improved your ability to manage financial aspects of your business?

4. **Have you experienced any social or cultural impacts as a result of utilizing digital platforms for your business?**
   - Probe: How have digital tools influenced perceptions of women entrepreneurs within your community or industry?
   - Probe: Have you encountered any challenges or opportunities related to gender stereotypes in your business interactions online?

5. **Can you discuss any specific instances in which digital technologies have helped you overcome conventional barriers to women's economic participation?**
   - Probe: How has access to mobile banking or online marketplaces helped you to engage in business activities that were inaccessible before?
   - Probe: Have digital platforms contributed to your business's growth or sustainability in ways that offline methods could not have done?

6. **How have digital platforms such as social media helped you expand your market reach and increase sales?**
   - Probe: Can you describe specific instances where social media platforms have enabled you to reach customers beyond your local market?
   - Probe: In what ways has digital marketing through social media platforms like Facebook or Instagram contributed to business growth?

7. **In what ways do mobile applications assist you in managing day-to-day operations and enhancing business efficiency?**
   - Probe: How do mobile apps for inventory management and customer communication streamline your business processes?
   - Probe: Have these apps helped you in improving customer satisfaction or reducing operational costs? Please provide examples.

8. **Could you elaborate on how access to mobile banking services has empowered you to manage finances and support business expansion?**
   - Probe: How has mobile banking facilitated access to credit or financial services that were previously challenging to obtain?
   - Probe: Can you share a specific instance where mobile banking has played a crucial role in your business decision-making or growth strategy?

9. **How have online forums or digital platforms facilitated networking and collaboration opportunities for your business?**
   - Probe: Have you connected with mentors or industry peers through online platforms? How has this networking benefited your business?
   - Probe: Can you outline some ways in which online communities contribute to knowledge sharing and professional development in your field?

10. **In what ways has access to the internet empowered you to make independent decisions and enhance your business acumen?**
    - Probe: Can you provide examples of how online resources have helped you stay informed about market trends or industry developments?
    - Probe: How do you utilize online information to negotiate contracts or make strategic decisions for your business?

11. **How have digital platforms challenged traditional gender norms and contributed to your recognition in the industry?**

    - Probe: Have digital platforms helped in showcasing your products or services in ways that challenge stereotypes in your industry?
    - Probe: Can you discuss any instances where digital visibility has positively impacted your business reputation or customer perception?

These questions were constructed to explore the unique ways in which digital platforms and tools have helped to empower women in Bangladesh, with specific focus on their experiences with market expansion, operational efficiency, financial management, networking, decision-making autonomy, and social empowerment. Adjust the wording further based on specific nuances or preferences of the interviewees to ensure clarity and relevance during the interviews.

3 Mobile Phones and the Internet as Catalysts for Women Empowerment in Developing Countries

A Systematic Review

D. Simona, K. S. Amritha, and B. Daniel

3.1 INTRODUCTION

The rapid proliferation of mobile and internet technologies globally has ushered in profound societal transformations, particularly in developing regions, where access to these digital tools has expanded exponentially. As of 2023, there are over 5 billion mobile phone users and nearly 5.3 billion internet users worldwide (Elagina, 2025). These technologies have played a pivotal role in advancing women's empowerment across various domains. Economically, mobile banking and microfinance applications have empowered women by providing access to financial services and facilitating independent financial management (Shanahan, 2022). Educationally, online platforms have revolutionized learning opportunities for women, offering flexibility in accessing education and skills development and, thereby, overcoming traditional barriers such as distance and societal norms (United Nations Educational, Scientific and Cultural Organization, 2021). Socially and politically, digital platforms, particularly social media, have become powerful tools for advocacy, enabling women to amplify their voices, mobilize communities, and advocate for gender equality, as seen in movements like #MeToo and #TimesUp (Mendes & Ringrose, 2019). Additionally, digital health initiatives have enhanced healthcare access for women in remote and underserved areas through telemedicine services and health information dissemination via mobile apps, contributing to improved maternal health and reproductive rights. Despite these advancements, significant challenges persist, including digital literacy gaps and unequal access to technology, particularly affecting women in rural and marginalized communities. Addressing these challenges is crucial to fully utilizing the advantages of digital technology for women's empowerment and achieving comprehensive gender equality goals.

 DOI: 10.1201/9781003544838-3

3.2 WOMEN EMPOWERMENT

Women empowerment is defined as a means of improving women's access to and control over resources, their capacity to make wise decisions in life, and their ability to confront discriminatory practices. It encompasses multifaceted strategies aimed at achieving gender equality and promoting inclusive growth across economic, educational, political, social, and legal domains. This chapter explores the transformative impact of women empowerment on individual empowerment and societal progress.

3.2.1 Economic Empowerment

Economic empowerment aids in evening access to economic opportunities, resources, and financial independence for women. This includes:

1. **Access to Education and Skills Training**: Education plays a crucial role in empowering women economically by equipping them with the needed skills and inputs to participate in the workforce and entrepreneurial activities. Studies have shown that education enhances individual earning potential, thus promoting economic growth and poverty reduction (World Bank, 2022).
2. **Job Opportunities and Entrepreneurship Support**: Providing women with equal access to employment opportunities and supporting female entrepreneurship are essential aspects of economic empowerment. Policies that promote women's access to credit, markets, and business networks can significantly enhance their businesses' economic outcomes (United Nations Entity for Gender Equality and the Empowerment of Women, 2023).
3. **Fair Wages and Labor Rights**: Ensuring fair wages and protecting women's labor rights are critical for economic empowerment. Closing gender pay gaps and promoting equal pay for equal work contribute to economic justice and empowerment.

3.2.2 Educational Empowerment

Educational empowerment focuses on providing girls and women with quality education and lifelong learning opportunities. This includes:

1. **Access to Quality Education**: Ensuring girls' and women's access to primary, secondary, and higher education is fundamental to their empowerment. Education enhances individual capabilities and contributes to improved health outcomes, reduced child mortality, and increased economic productivity (United Nations Educational, Scientific and Cultural Organization, 2020).
2. **Skills Development and Vocational Training**: Offering vocational training and skills development programs customized to women's requirements enhances their employability and economic independence. Skills training programs that focus on non-traditional fields can help break gender stereotypes and empower women in diverse sectors (United Nations Entity for Gender Equality and the Empowerment of Women, 2023).

3. **Promoting Literacy and Lifelong Learning**: Literacy programs and adult education initiatives are crucial for empowering women who have been historically marginalized from formal education systems. Lifelong learning opportunities enable women to adapt to changing economic and social environments and participate more effectively in community development (United Nations Educational, Scientific and Cultural Organization, 2020).

3.2.3 Political Empowerment

Political empowerment aims to promote women's involvement and leadership in political processes and organizations. This includes:

1. **Representation in Decision-Making**: Increasing women's representation in elected offices, government positions, and decision-making bodies is crucial for ensuring that women's voices and interests are addressed and integrated into policy-making processes (Inter-Parliamentary Union, 2021).
2. **Advocacy for Gender-Responsive Policies**: Advocating for policies that focus on gender-based discrimination, enhance women's rights, and ensure equal opportunities in political and public life is essential for political empowerment. Gender mainstreaming strategies help integrate gender perspectives into all policies and programs (United Nations Entity for Gender Equality and the Empowerment of Women, 2023).
3. **Building Leadership Capacities**: Providing training and mentorship programs that build women's leadership skills and confidence in political arenas enhances their ability to effectively advocate for gender equality and drive social change (Inter-Parliamentary Union, 2021).

3.2.4 Social Empowerment

Social empowerment involves challenging harmful social norms and stereotypes that perpetuate gender inequality. This includes:

1. **Promoting Gender Equality Awareness**: Raising awareness about gender equality issues and promoting positive attitudes toward gender equality within families, communities, and institutions are essential for social empowerment. Education and advocacy campaigns can help challenge stereotypes and discriminatory practices.
2. **Addressing Violence against Women**: To protect women's safety and well-being, it is essential to support programs that prevent and address violence against women, such as sexual harassment, domestic abuse, and human trafficking. Legal reforms and support services for survivors are essential components of efforts to empower women socially (United Nations Entity for Gender Equality and the Empowerment of Women, 2023).
3. **Fostering Inclusive Communities**: Creating inclusive communities where women have equal access to social services, cultural activities, and public spaces promotes their social integration and participation. Community-based

initiatives that empower women economically and politically also contribute to social cohesion and development.

3.2.5 Legal Empowerment

Legal empowerment focuses on ensuring women's rights are protected and upheld through legal reforms and enforcement mechanisms. This includes:

1. **Ensuring Legal Rights**: Advocating for and enforcing laws that protect women's rights to property ownership, inheritance, and access to justice are crucial for legal empowerment. Legal reforms that eliminate discriminatory practices and promote gender equality in the legal system contribute to women's empowerment.
2. **Access to Justice**: Ensuring women have access to affordable and effective legal services and mechanisms for resolving disputes, including family and employment-related issues, enhances their ability to exercise their rights and seek redress for violations (United Nations Entity for Gender Equality and the Empowerment of Women, 2023).
3. **Combating Discriminatory Laws**: Addressing discriminatory laws and policies that perpetuate gender-based violence, discrimination, and inequality is essential for legal empowerment. Advocacy efforts and international human rights frameworks provide mechanisms for challenging and reforming unjust laws.

3.3 THEORETICAL FRAMEWORK

The **agency theory** suggests that individuals have the potential (agency) to come up with decisions and take actions that affect their lives. In the context of technology and women empowerment, this theory examines how access to mobile phones and the internet can empower women by providing them with tools to access information, communicate with others, and participate in economic and social activities (Bandura, 2001). Women who have access to mobile phones can use them to access educational resources, job opportunities, healthcare information, and financial services. In addition to improving their social networks and support systems, mobile phone communication can provide them with more agency and decision-making authority in their homes and communities.

The **digital divide theory** explores inequalities in access to and use of digital technologies, such as mobile phones and the internet, across different demographic groups, including women (Warschauer, 2003). In many societies, women face barriers such as affordability, literacy, cultural norms, and restrictive gender roles that limit their access to digital technologies. This digital divide can perpetuate existing gender inequalities by limiting women's opportunities for education, economic participation, and social engagement. Bridging the digital divide through policies and initiatives that promote equal access to technology can help empower women by enabling them to access information, resources, and opportunities.

The **capability approach**, developed by Amartya Sen and Martha Nussbaum, focuses on individuals' capabilities—their opportunities to choose and achieve well-being. This strategy highlights how having access to mobile phones and the internet may increase women's access to social networks, healthcare, education, and employment prospects in the context of technology and women's empowerment (Sen, 1999). For instance, mobile phones aid women in rural areas in accessing agricultural information, market prices, and financial services, thereby enhancing their economic capabilities. Internet access can provide women with opportunities for online education and skills development, enhancing their capabilities to pursue higher-paying jobs or entrepreneurial ventures.

Feminist technoscience examines how gender norms and power dynamics influence the design, development, and use of technologies. It emphasizes the importance of considering gender perspectives in technology adoption and implementation to promote women's empowerment (Wajcman, 2007). Technologies that uphold conventional gender norms or exclude women from technological development decision-making processes are criticized by this approach. It advocates for inclusive and gender-sensitive approaches to technology design and policy-making, ensuring that women's voices and needs are represented. Feminist technoscience seeks to empower women by addressing gender inequities in technology and offering them tools that improve their chances, safety, and autonomy.

Intersectionality, proposed by Kimberlé Crenshaw, investigates the ways in which many social categories—such as gender, color, class, etc.—intersect to influence people's opportunities and experiences. Intersectionality aids in understanding how variables other than gender, such as socioeconomic class, ethnicity, location, and handicap status, impact women's access to and advantages from mobile and internet technologies in the context of technology and women's empowerment (Crenshaw, 1991). For instance, women from marginalized communities may face compounded barriers to accessing digital technologies due to poverty, discrimination, or lack of infrastructure in their regions. Intersectionality calls for policies and interventions that address these intersecting inequalities to ensure that all women, regardless of their backgrounds, can benefit from technology and participate fully in economic, social, and political life.

3.4 RATIONALE OF THE STUDY

The study on the impact of mobile and internet penetration on women empowerment is essential due to its potential to catalyze women's socioeconomic advancement and societal inclusion. Understanding how increased access to mobile and internet technologies enhances women's economic opportunities, social participation, and health outcomes can inform policies and interventions aimed at fostering gender equality. This research also sheds light on the transformative role of digital connectivity in challenging traditional gender roles and promoting women's leadership in digital economies, thereby contributing to broader discussions on empowerment and societal progress.

The existing literature consistently highlights that increased access to mobile phones and the internet correlates positively with enhanced economic opportunities for women, particularly in developing regions (GSMA, 2020). Mobile banking services, for instance, have empowered women by providing access to financial services that were previously inaccessible due to geographical and societal barriers (International Telecommunication Union, 2022). Moreover, digital connectivity has facilitated women's involvement in decision-making processes and entrepreneurial activities, contributing to their economic independence and social mobility (United Nations Educational, Scientific and Cultural Organization, 2020). However, challenges such as digital literacy gaps and affordability constraints continue to hinder equitable access to these technologies, especially among marginalized groups (World Bank, 2021). Therefore, while acknowledging the substantial benefits, further research is needed to explore effective strategies for ensuring inclusive digital empowerment that reaches all women, irrespective of socioeconomic status or geographic location.

3.5 METHODOLOGY

3.5.1 Eligibility Criteria

The eligibility criteria were established to ensure the inclusion of relevant studies focused on the role of mobile phones and the internet in empowering women in developing countries. The studies had to meet the following requirements:

1. **Population:** Women in developing countries.
2. **Interventions:** Use of mobile phones and internet-based technologies.
3. **Outcome:** Evidence of empowerment in terms of economic, social, and political aspects.
4. **Study Design:** Quantitative, qualitative, or mixed-method studies.
5. **Language:** Studies published in English were included.
6. **Publication Date:** Only studies published between 2000 and 2024 were considered.

3.5.2 Information Sources

An extensive search was conducted across multiple platforms, including PubMed, Scopus, Web of Science, Google Scholar, and other relevant academic sources. Grey literature, such as reports from NGOs and government publications, was also considered.

3.5.3 Search Strategy

The search strategy involved combining keywords and terms related to women's empowerment, mobile phones, the internet, and developing countries. The search queries included terms such as:

1. "Women empowerment" AND "mobile phones" AND "developing countries"
2. "Internet access" AND "women's rights" AND "low-income countries"
3. "Digital inclusion" AND "women" AND "global south"

Boolean operators (AND, OR) were used to ensure a wide but focused retrieval of studies.

3.5.4 Study Selection Process

The selection process followed the PRISMA guidelines for transparency. The studies were screened in three phases:

1. **Initial Screening:** Titles and abstracts were reviewed for relevance.
2. **Full-Text Screening:** The full text of potentially relevant studies was retrieved and assessed for eligibility based on the established criteria.
3. **Final Inclusion:** Studies that met all criteria were included in the review.

3.5.5 Data Extraction Process

Data from the included studies were extracted systematically using a standardized extraction form. The extracted data included author(s), year of publication, study design, sample size and characteristics. intervention type (mobile phones, internet), country of study, key findings related to women's empowerment, and limitations of the study.

3.5.6 Synthesis of Results

The data from the selected studies were synthesized narratively and, where possible, quantitatively. The studies were grouped by themes such as:

1. Economic empowerment through mobile phones (e.g., entrepreneurship, access to financial services).
2. Social empowerment via internet use (e.g., education, healthcare access).
3. Political empowerment through digital inclusion (e.g., civic participation).

A meta-analysis was not feasible due to the heterogeneity of the study designs and outcomes. Instead, thematic analysis was conducted to identify common patterns and trends across the studies.

3.5.7 Ethical Considerations

Since this is a systematic review, no primary data were collected, and thus, no ethical approval was required. However, ethical standards of transparency and rigor in the synthesis of existing literature were adhered to throughout the process.

3.5.8 Findings

3.5.8.1 Challenges and Barriers

1. **Gender-specific barriers to access to and use of technology:**
   - **Digital Literacy**: Women often experience disparities in digital literacy skills compared to men, which affects their ability to effectively use and

benefit from technology. According to the World Bank, gender gaps in digital skills persist globally, with women often lagging behind in acquiring the necessary digital literacy for economic and social empowerment (World Bank, 2021). This gap limits women's participation in digital economies and their access to online educational resources and job opportunities.

- **Access to Devices**: Economic disparities and cultural norms contribute to unequal access to smartphones and internet-enabled devices among women. According to GSMA research, women are less likely than males to own a mobile phone in many low- and middle-income nations, which limits their access to digital services like social networks, mobile banking, and healthcare information (GSMA, 2020). This disparity perpetuates gender inequalities in accessing opportunities facilitated by digital technologies.
- **Control and Ownership**: Societal norms and patriarchal structures often limit women's control over technology use, including decision-making regarding mobile phone ownership and usage. UN Women highlights that in some cultures, women have limited autonomy with regard to managing their own digital devices, which can hinder their ability to leverage technology for personal and economic empowerment (United Nations Entity for Gender Equality and the Empowerment of Women, 2019). Addressing these structural barriers is crucial for enabling women to use technology to improve their livelihoods and participate fully in digital societies.

2. **Cultural and societal challenges:**
    - **Norms and Values**: Norms and values shape women's access to and use of technology, influencing patterns of digital involvement. Research shows that gender influences how individuals perceive and adopt communication technologies, with women often reporting lower access and usage due to cultural barriers and social norms prioritizing men's technology use for education, entrepreneurship, and networking (Hilbert, 2011).
    - **Gender Roles**: Traditional gender roles often assign men as primary users and decision-makers regarding technology within households. The Pew Research Center highlights how these gendered expectations impact technology adoption rates, with men often having greater access to and control over digital devices than women (Pew Research Center, 2021). This disparity limits women's opportunities to develop digital skills and participate in online communities where technology plays a significant role in economic and social interactions.
    - **Privacy Concerns**: Privacy concerns and social norms can deter women from using mobile phones and the internet, particularly in contexts where cultural expectations regarding women's behavior and communication are restrictive. The International Telecommunication Union (ITU) reports on how cultural norms and privacy concerns influence women's digital behaviors and access to online information and services (International Telecommunication Union, 2022). Addressing these concerns requires promoting digital literacy and creating safe online environments that respect women's privacy and autonomy.

3. **Technological barriers:**
   - **Infrastructure**: Unequal distribution of network coverage and unreliable electricity supply in underdeveloped regions limit women's access to mobile and internet services. The Alliance for Affordable Internet highlights infrastructure challenges as a significant barrier to women's digital inclusion, particularly in regions where basic infrastructure development lags behind that in urban centers (Alliance for Affordable Internet, 2021). Improving infrastructure accessibility is essential to ensuring equitable digital opportunities for women in remote and marginalized communities.
   - **Affordability**: High costs associated with data plans, internet-enabled devices, and digital services pose financial barriers for women, especially those from low-income households. The United Nations Development Programme (UNDP) emphasizes affordability as a critical factor influencing women's ability to benefit from digital technologies for economic empowerment (United Nations Development Programme, 2020). Women's access to digital resources and opportunities can be improved by addressing affordability through community initiatives, subsidized programs, and legislative actions.
   - **Content Relevance**: Availability of culturally relevant online content is essential for engaging women in digital platforms and maximizing the empowerment impacts of technology. Evidence suggests that digital literacy training tailored to women's needs can improve online engagement, perceived ICT competence, and participation in digital opportunities such as education and employment (Mukherjee et al., 2019). When digital content resonates with women's lived experiences and priorities, stakeholders are more likely to see greater participation and engagement in digital spaces.

3.5.9 Impact on Education

3.5.9.1 E-Learning Platforms and Skills Development

1. **Accessibility and Flexibility**: E-learning platforms have revolutionized access to education by overcoming traditional barriers such as geographical location and time constraints. Women, particularly those in rural or underserved areas, can now access a wide range of courses and educational resources online. Platforms like Coursera, Khan Academy, and Udemy offer diverse courses, from technical skills like programming and data science to soft skills like leadership and communication (United Nations Educational, Scientific and Cultural Organization, 2020). Because of this accessibility, women can learn new skills or improve their current ones from the comfort of their homes, allowing them to fit education around their obligations to their families and their jobs.
2. **Skill Enhancement and Career Advancement**: E-learning enables women to develop technical competencies crucial for career advancement in fields traditionally composed of men, such as STEM (Science, Technology, Engineering, and Mathematics). For instance, initiatives like TechWomen provide mentorship and e-learning opportunities specifically tailored to women

in STEM, fostering leadership and entrepreneurship skills. By gaining proficiency in digital technologies and specialized knowledge areas, women can compete on a more equal footing in the job market and pursue leadership roles in their respective fields.

3.5.9.2 Literacy Rates and Education Attainment among Women

1. **Improving Literacy Rates**: E-learning programs targeted at adult learners, including women, have proven effective in improving literacy rates, particularly in developing countries. These programs often utilize mobile-based apps and interactive platforms that cater to diverse learning needs and levels of literacy. For example, literacy apps designed with user-friendly interfaces and audiovisual content help women in marginalized communities develop basic literacy skills in their local languages (United Nations Educational, Scientific and Cultural Organization, 2020). Such initiatives not only enhance women's ability to read and write but also empower them to engage more actively in social and economic activities.
2. **Closing Gender Gaps in Education**: Technology-enabled educational interventions play a crucial role in narrowing gender disparities in education attainment. By providing equitable access to quality learning resources, e-learning helps bridge the gap between male and female students. Projects like the One Laptop Per Child initiative have distributed laptops and tablets to schoolgirls in rural areas, contributing to improved learning outcomes and retention rates. These initiatives are instrumental in ensuring that women have even opportunities to acquire the education and skills necessary for personal development and economic empowerment.

3.5.10 Impact on Health

3.5.10.1 Mobile Health (mHealth) Applications and Services

1. **Accessibility and Convenience**: Mobile health applications provide convenient access to healthcare information and services anytime and anywhere through smartphones and other mobile devices. These apps offer functionalities such as symptom checkers, telemedicine consultations, medication reminders, and health-monitoring tools. For women, especially those in remote or underserved areas, mHealth overcomes geographical barriers, enabling them to receive timely medical advice and support without needing to travel long distances.
2. **Health Monitoring and Management**: mHealth apps empower women to monitor their health conditions and manage chronic diseases more effectively. For example, apps for maternal health track pregnancy milestones, provide prenatal care information, and offer support for postpartum care. Similarly, diabetes management apps help women to track blood glucose levels and manage medication adherence, improving overall health outcomes (United Nations Development Programme, 2020). These tools promote proactive healthcare management and empower women to take charge of their health.

3.5.10.2 Access to Healthcare Information and Services for Women

1. **Health Education and Awareness**: mHealth platforms serve as valuable tools for health education and awareness campaigns targeted at women. These platforms deliver culturally sensitive information on reproductive health, family planning, nutrition, and disease prevention through text messages, videos, and interactive content (United Nations Entity for Gender Equality and the Empowerment of Women, 2018). By increasing health literacy, women can be enabled to make informed decisions about their healthcare needs and practices, leading to improved maternal and child health outcomes.
2. **Telemedicine and Remote Consultations**: Telemedicine services offered through mHealth apps connect women with healthcare providers remotely, overcoming barriers such as limited access to healthcare facilities or transportation challenges. Virtual consultations enable timely diagnosis, treatment, and monitoring of health conditions, particularly in rural or marginalized communities (International Telecommunication Union, 2022). This approach enhances continuity of care and reduces disparities in healthcare access for women in underserved areas.

3.5.11 Impact on Social Inclusion

3.5.11.1 Connectivity, Networking, and Community-Building

1. **Enhanced Connectivity**: Technology facilitates broader social connectivity by enabling women to engage in virtual communities, social networks, and online forums. Social media platforms such as Facebook, Twitter, and Instagram provide spaces for women to connect, share experiences, and build supportive networks locally and globally (United Nations Educational, Scientific and Cultural Organization, 2020). This connectivity reduces isolation, fosters collaboration, and amplifies voices that may otherwise be marginalized in traditional offline spaces.
2. **Networking and Support Groups**: Online platforms host networking opportunities and support groups specifically tailored to women's needs and interests. For instance, professional networking sites like LinkedIn offer career development resources and mentorship programs that empower women in advancing their careers and entrepreneurial endeavors. Similarly, online forums and peer support groups provide emotional support, advocacy, and resources for women facing various challenges, from health issues to workplace discrimination.

3.5.11.2 Civic Participation and Empowerment

1. **Political Engagement**: Digital technologies facilitate women's engagement in civic and political processes by providing platforms for advocacy, information- sharing, and mobilization. Social media campaigns and digital activism movements have empowered women to raise awareness about gender-based violence, advocate for policy reforms, and participate in elections as candidates or informed voters (United Nations Entity for Gender Equality and the Empowerment of Women, 2021). These platforms amplify women's

voices in the public discourse and contribute to shaping inclusive and equitable societies.

2. **Educational and Economic Empowerment**: Online education and digital literacy initiatives empower women with knowledge and skills that enhance their economic opportunities and decision-making capabilities. Access to online courses, webinars, and e-learning platforms equips women with skills in entrepreneurship, financial management, and leadership, enabling them to contribute more effectively to economic development and community welfare (World Bank, 2019). This educational empowerment fosters independence and resilience, reducing economic disparities and enhancing social inclusion.

3.6 CONCLUSION

This systematic review demonstrates the transformative significance of mobile phones and the internet in advancing women's empowerment across developing countries. The findings indicate that digital technologies significantly impact women's social, economic, and political empowerment by enhancing access to information, fostering educational opportunities, and creating new avenues for economic independence. Mobile phones and internet access have provided women with tools for communication, entrepreneurship, and participation in social and political dialogues which were previously inaccessible or limited due to geographical, cultural, and economic barriers.

However, while the positive impacts are evident, the review also highlights persistent challenges, including digital literacy gaps, gender-based digital exclusion, and socio cultural constraints that continue to hinder full empowerment. These barriers suggest that while technology is a critical enabler, its benefits cannot be fully realized without addressing the underlying inequalities in education, income, and societal norms that disproportionately affect women.

To further accelerate progress, there is a need for targeted policies and initiatives that promote equal access to digital technologies and encourage inclusive digital education and training. Investments in infrastructure, especially in rural and underserved areas, alongside efforts to dismantle societal stereotypes, will be crucial for bridging the digital divide. Moreover, partnerships between governments, the private sector, and civil society can foster innovation and support sustainable models of digital inclusion that specifically cater to women's needs.

In conclusion, mobile phones and the internet represent powerful tools for women's empowerment, but realizing their full potential requires a multifaceted approach that combines technology with broader efforts to challenge and overcome entrenched gender disparities. With continued attention to these areas, developing countries can harness digital innovations to build more equitable and inclusive societies for women.

REFERENCES

Alliance for Affordable Internet. (2021). The mobile broadband pricing landscape.

Bandura, A. (2001). Social cognitive theory: An agentic perspective. *Annual Review of Psychology, 52*(1), 1–26. https://doi.org/10.1146/annurev.psych.52.1.1

Crenshaw, K. (1991). Mapping the margins: Intersectionality, identity politics, and violence against women of color. *Stanford Law Review*, *43*(6), 1241–1299.
Elagina, D. (2025, November 19). Global smartphone penetration 2016-2023. *Statista*. www.statista.com/statistics/203734/global-smartphone-penetration-per-capita-since-2005/
GSMA. (2020). Connected women: The mobile gender gap report 2020. https://www.gsma.com/mobilefordevelopment/resources/mobile-gender-gap-report-2020/
Hilbert, M. (2011). Digital gender divide or technologically empowered women in developing countries? A typical case of lies, damned lies, and statistics. *Women's Studies International Forum*, *34*(6), 479–489. https://doi.org/10.1016/j.wsif.2011.07.001
International Telecommunication Union. (2022). Measuring digital development: Facts and figures 2022. ITU. https://www.itu.int/en/mediacentre/backgrounders/Pages/bridging-the-gender-divide.aspx
Inter-Parliamentary Union. (2021). Women in parliament in 2021. IPU. https://www.ipu.org
Mendes, K., & Ringrose, J. (2019). Digital feminist activism: #MeToo and the everyday experiences of challenging rape culture. In B. Fileborn & R. Loney-Howes (Eds.), *#MeToo and the Politics of Social Change* (pp. 37–51). Springer. https://doi.org/10.1007/978-3-030-15213-0_3
Mukherjee, T., Ilavarasan, P. V., & Kar, A. K. (2019, January). Digital literacy training, impact & moderating role of perceived value among unemployed women in India. In *Proceedings of the Tenth International Conference on Information and Communication Technologies and Development* (pp. 1–4). https://doi.org/10.1145/3287098.3291932
Pew Research Center. (2021). Mobile fact sheet. Pew Research Center. https://www.pewresearch.org
Sen, A. (1999). *Development as Freedom*. New York, NY: Alfred A. Knopf.
Shanahan, M. (2022, June 22). The mobile gender gap report 2022. Mobile for Development. www.gsma.com/mobilefordevelopment/blog/the-mobile-gender-gap-report-2022/
United Nations Development Programme. (2020). Human development report 2020: The next frontier—Human development and the Anthropocene. UNDP.
United Nations Educational, Scientific and Cultural Organization. (2020). Global education monitoring report 2020: Inclusion and education—All means all. UNESCO.
United Nations Educational, Scientific and Cultural Organization. (2021). Technology in education: A tool on whose terms? UNESCO.
United Nations Entity for Gender Equality and the Empowerment of Women. (2018). Turning promises into action: Gender equality in the 2030 Agenda for Sustainable Development. UN Women.
United Nations Entity for Gender Equality and the Empowerment of Women. (2019). Progress of the world's women 2019–2020: Families in a changing world. UN Women.
United Nations Entity for Gender Equality and the Empowerment of Women. (2021). Beyond COVID-19: A feminist plan for sustainability and social justice. UN Women.
United Nations Entity for Gender Equality and the Empowerment of Women. (2023). Digital innovation for gender equality. UN Women.
Wajcman, J. (2007). From women and technology to gendered technoscience. *Information, Communication & Society*, *10*(3), 287–299.
Warschauer, M. (2003). *Technology and Social Inclusion: Rethinking the Digital Divide*. Cambridge, MA: MIT Press.
World Bank. (2019). World development report 2019: The changing nature of work. World Bank.
World Bank. (2021). Gender and inclusion in digital development. World Bank. https://www.worldbank.org/en/topic/digital/brief/gender-and-inclusion-in-digital
World Bank. (2022). Bridging the gender digital divide: Final report. World Bank.

4 A Conceptual Framework for Women's Entrepreneurship Policy Aligning Government and Economic Objectives

Mohammed Imran, Alok Ranjan Behera, and Sabina Begum

4.1 INTRODUCTION

Any woman or women-led organisation that starts, manages, and grows a company is considered an entrepreneur. An enterprise is considered women-owned and controlled if women own and control at least 51 per cent of the capital and provide at least 51 per cent of the employment within the enterprise. This definition is provided by the Government of India. Several forces at work empower women to strike out on their own and start their businesses. The driving force behind this behaviour is a strong desire to make autonomous decisions regarding one's life and professional trajectory. Women long for freedom from the constraints of domestic duties and tasks. As a result of these influences, women entrepreneurs often pick a career path that presents both intellectual and personal challenges. Pull factors are what we call these kinds of situations. Women are involved in business activities as a result of familial pressure and the burden of responsibility, which are examples of push forces (Goyal & Parkash, 2011).

To be an entrepreneur is to plan, create, and manage a new firm from the ground up, whether it be a service, product, or process. Entrepreneurship is defined as the ability and desire to create, plan, and oversee a commercial enterprise while bearing the risk of that endeavour in order to turn a profit (Pallavi et al., 2022).

For every entrepreneur, the *Bhagavad Gita* offers a treasure trove of wisdom on how to succeed in life, in leadership roles, and in business without sacrificing one's ideals. "Guru Brahma Guru Vishnu, Gurur Devo Maheshwaraha, Guru Saakshat Para Brahma, Tasmai Shree Guruve Namaha" summarises the central idea of leadership contained in the book. A Guru is truly an incarnation of the trinity of Brahma (the creator), Vishnu (the sustainer and saviour), and Shiva (the destroyer). Through a

DOI: 10.1201/9781003544838-4

guru, wisdom is born, knowledge is nurtured, and ignorance is eradicated. A person's Guru deserves respect from all. Young entrepreneurs are thriving in today's market; what's lacking, though, is the advice and instruction of seasoned "gurus" on how to succeed. A person's enthusiasm for making their dreams a reality is a common denominator among entrepreneurs. If the circumstances are right, it could be born at any moment and flourish. Living your passion is the first stage of being an entrepreneur, but doing it right strategically is the second stage. At this point, the business and the entrepreneur's ability to lead are defined.

Educated women do not like to confine themselves to the confines of the home. When they're with someone, they expect the same level of respect. There is a long way to go before Indian women achieve equal rights and positions due to the deep-seated traditions in Indian culture and the historically male-dominated sociological framework. Throughout their lives, women are expected to rely on men, both within and outside of the home, due to the widespread belief that women are weaker than men. Women in Indian culture were relegated to the role of mere executors, carrying out the will of the men in their families. Women may be the most underappreciated asset in the world, even though they constitute half of the world's brainpower. Several inspiring women entrepreneurs in India overcame societal obstacles to achieve great success. They are praised for their accomplishments in their profession and stand out from the rest of the crowd. As a result of women's rising educational status and other desires for better living, the social fabric of Indian society has changed, and with it, so has the way Indian women live their lives. Indian women have always competed with men and triumphed over them, and the economic world is no different. These women leaders are not afraid to speak their minds or take calculated risks. Thanks to their determination, hard effort, and endurance, they were able to survive and thrive in this extremely competitive environment. Indian women business owners excel because they are persuasive, have an open approach to solving problems, aren't afraid to take chances, can inspire others, and know how to win and lose graciously (Goyal & Parkash, 2011).The purpose of this research is to examine the initiatives and policies that support female entrepreneurs in India.

4.2 THE DEVELOPING COUNTRIES SETTING FOR WOMEN ENTREPRENEURS

Delivering high-quality goods and services isn't enough to succeed in today's cutthroat global marketplace. Entrepreneurs understand that their products and services, no matter how high-quality they are, will probably become commodities in a short amount of time. More options and more information than ever before are available to shoppers in today's global online marketplace. It gives individuals the freedom to shop the way they want to, based on their priorities—whether that's price, service, brand, or something else entirely. The experience that a company provides to its clients is increasingly the defining factor among competing businesses. Businesses that put their customers first will examine and improve every aspect of the customer journey, from the people involved to the systems they use. By centring operations around the customer, businesses can maximise their impact on customer satisfaction and, by extension, their bottom line.

A firm can't get off the ground without a financially sound concept and sufficient funding. At this point, because there is so much rivalry in the market, it is impossible to triumph over it. Marketers have grown too big, consumers want more options, capital is plentiful and ready to be invested, fresh opportunities will attract everyone and their dog, and consumers' preference cycles are shorter; these and many more reasons are at play. However, to stay alive, one can find a way to give the competition the boot in the hopes that it won't come back. Every business owner needs to be on the lookout for these recurrences and to be prepared to get back into action whenever they happen.

India has a low rate of women entrepreneurship, ranking 52nd out of the 57 nations surveyed for women entrepreneurship. Women make up 13.76 per cent of India's overall entrepreneurial population, with 8.05 million out of 58.5 million firms being owned by women. However, India ranks 52nd out of 57 nations in economic participation and opportunity. Women are under-represented in positions of power and authority in business, professional fields, and technical occupations and are less likely to join the labour force. The percentage of women in the labour force has dropped from 35 per cent in 2005 to 26 per cent in 2018, despite strong economic development. States with better literacy rates also have more women entrepreneurs, with Tamil Nadu, Kerala, Andhra Pradesh, West Bengal, and Maharashtra having the highest proportion of women entrepreneurs. Women's business practices in India are varied, with 98 per cent of women-owned firms being micro-enterprises, largely attributable to microfinance (Subbarayudu & Rao, 2021).

The moment has come to acknowledge that men were once the dominant social group. The woman's dream of starting a business was just a pipe dream when she had to perform all the housekeeping first. The tides have turned, and there is an abundance of strong women who have grown beyond many types of oppression.

Over the past several years, India's start-up ecosystem has seen steady growth and development. The entrepreneurial sectors of India are thriving, and women are playing an equally important role to that of men. Start-ups and the idea of being one's own boss have captured the imagination of many, including many women in India. It may come as a surprise to hear that women make up 13.76 per cent of India's business owners according to a poll conducted by the National Statistics and Programme Implementation Department. Women excel in many areas, both at home and in the workplace, which is well-known. And it is precisely because of this trait that they can launch prosperous businesses. The task of turning an entrepreneur's dreams into a reality isn't easy, but everyone has to start somewhere. These women achieved their level of business success by doing just that. Instead of limiting themselves to "what ifs," they simply began. To help India's economy thrive, we've compiled a list of the top 10 women entrepreneurs in the country.

The number of women entrepreneurs in India, under the Prime Minister Employment Generation initiative, increased from 14,685 in 2010–2011 to 19,104 in 2019–2020 (Figure 4.1).

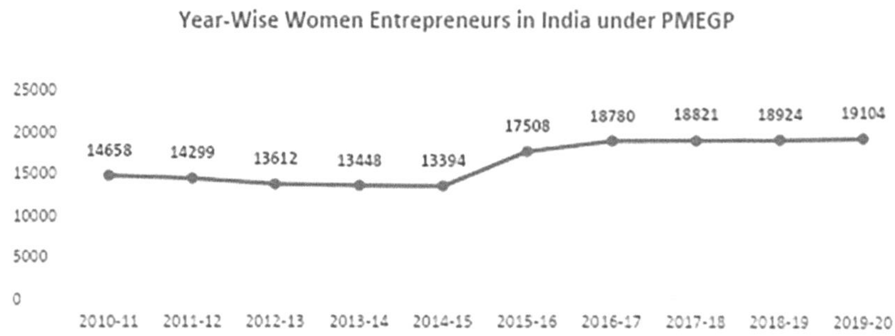

FIGURE 4.1 Year-Wise Women Entrepreneurs in India under PMEGP.

Source: Ministry of MSME 2020.

4.3 JUSTIFICATION FOR DIVERSITY

Equally important as male entrepreneurs are women, who take risks, organise, and innovate. Even though they are considered the superior half of society, Indian women nevertheless face discrimination and inequality. The majority of the variation in entrepreneurial styles between the sexes is attributable to perceptual characteristics, which significantly impact women's entrepreneurial inclination. Women face social disadvantages because of their lower literacy rate (40 per cent) and workforce participation rate (12 per cent), compared to men, whose literacy, workforce participation, and entrepreneurial engagement rates stand at 60 per cent, 52 per cent, and 18 per cent, respectively. Here are the findings from the survey carried out by IIT, Delhi (Sharma, 2013).

1. In the United States and Canada, women control a third of small businesses, and by the year 2020, that share will have jumped to half.
2. The proportion of working-age women in Asian countries is 40 per cent.
3. In China, women outnumber men by a margin of at least two lines.
4. Despite a rise from 7.69 per cent in 1992–1993 to 10 per cent in 2000–2001, the percentage of women business owners is still far lower than the general labour participation rate of 25.7 per cent.
5. There has been a meteoric surge in the representation of women in engineering, technical, and professional fields. Fewer than 15 per cent of the students enrolled in polytechnics and IITs go on to start their own businesses.

4.4 WOMEN ENTREPRENEURS AS A GLOBAL PHENOMENON

One of the highest percentage of women entrepreneurs is in Guatemala (Low Income), where it stands at 28.2 per cent. This suggests that women in this income group are very active entrepreneurs. Equally high, at 25.4 per cent, is Togo (Low Income), which may indicate that there are many women business owners in Togo and other low-income nations. Women also have a high rate of entrepreneurial activity in Chile, a middle-income country with a rate of 25.2 per cent. The United States of America has a somewhat high number of women entrepreneurs, ranking, with 18.1 per cent,

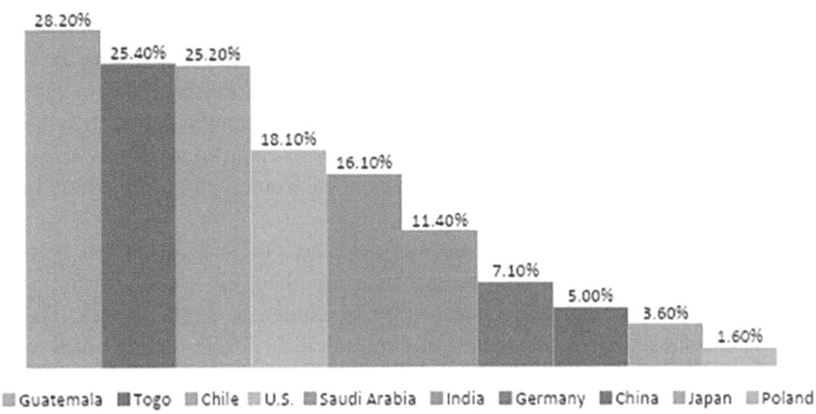

FIGURE 4.2 India and China Register Less Women Entrepreneurs.

Source: Global Entrepreneurship Monitor 2022.

among high-income countries. Women entrepreneurship is noteworthy in Saudi Arabia (High Income)—16.1 per cent. The amount of engagement may be influenced by socio-cultural elements in this high-income country. When compared to other countries in the same economic bracket, India's 11.4 per cent women entrepreneurship rate is lower, despite the country's status as a low-income one. Among high-income nations, Germany has the lowest rate of women entrepreneurship at 7.1 per cent. There is 5 per cent women entrepreneurships in China as shown in Figure 4.2, which is lower than the rates in other middle-income countries. Among Poles with middle-class incomes, the percentage of women entrepreneurs is rather low at 3.6 per cent. Japan (High Income) has a rate of 1.6 per cent; the lowest percentage of women entrepreneurs is found in Japan, which may indicate that there are cultural and economic obstacles there that prevent women from becoming business owners.

4.4.1 Methodology

This research relies on secondary data culled from a variety of sources, including the Bangladesh Bank, the Ministry of Micro, Small, and Medium Enterprises, the Economic Survey, scholarly journals, and reports. Quantitative and qualitative data are also covered in the study, along with descriptive data presented in the form of charts, bar graphs, and line diagrams.

4.4.2 Common Problems Faced by Bangladeshi and Indian Women Entrepreneurs

1. Relations within the family: Their families hold a special place in the hearts of Indian women. All of the housework, including taking care of the children and other family members, is intended to be their responsibility. Their time and energy are completely consumed by family obligations, such as providing

additional care to their husbands, children, and in-laws. Concentration and efficient business management will be next to impossible in such a setting.

2. Masculinity in the community: Male chauvinism persists despite the fact that our constitution guarantees gender equality. Unfortunately, women still face discrimination. They need the blessing of the family patriarch before they may join the professional world. The entrepreneurial spirit has always been associated with men. The advancement of women entrepreneurs is impeded by all of these factors.
3. Social obstacles: Traditional attitudes and practices in Indian civilisation regarding women might be a barrier to their advancement and success. Religion and caste coexist and impede women business owners. They have additional societal obstacles in rural regions. Their communities seem to always be watching them with an air of suspicion.
4. Unmet educational needs: When compared to males, women in India face a significant educational gap. A whopping 60 per cent of females fall into the no-diploma category. In comparison to educated men, educated women often acquire less or insufficient education due to factors such as poverty, early marriage, and the fact that sons tend to have greater educational opportunities. Women business owners still don't know much about government incentives, new technologies, production processes, marketing, and other innovations that can help them succeed since they haven't had the right education.
5. Issue with funds: When compared to male entrepreneurs, women entrepreneurs face more obstacles when trying to secure funding for their businesses. Women borrowers face a higher risk of company failure and lower credit-worthiness; that's why lending them money is something financial institutions, creditors, and banks are wary of. Customer payments that are either late or non-existent, in addition to financial issues with raw materials, work in progress, and finished goods, make matters worse.
6. Robust contests: The production process is often characterised by the use of low technology by women entrepreneurs. Fighting for survival in an overly competitive market against the organised sector and male counterparts who possess extensive expertise and the ability to implement cutting-edge technologies in business management is no easy feat.
7. Disinclination towards taking risks: Women business owners worry that they lack the necessary entrepreneurship skills. They aren't entrepreneurial at heart. There are still obstacles that women business owners face in the workplace, even after completing numerous entrepreneurship training programmes.
8. Skill gap in management: Only the most competent managers are able to carry out the increasingly specialised duties of management. Managerial tasks such as planning, organising, regulating, coordinating, staffing, leading, encouraging, etc. are not efficiently performed by women entrepreneurs. As a result, women's lack of managerial experience is preventing them from effectively running their businesses.

9. Feeling insecure: Women entrepreneurs have unique challenges due to their innate lack of self-confidence, which can be a driving force behind the success of their ventures. Finding a happy medium between running a family and a business is no easy feat. Striking a balance between the two often requires women to put their business spirit on hold.

4.4.3 Effects on Entrepreneurship by Women

1. Push factors: These include a lack of financial resources and discontent with current employment conditions. For instance, push factors can include difficulties in obtaining employment and the need for a more flexible work schedule so as to meet family obligations. To a greater extent than to males, these considerations may matter to women.
2. Pull factors: Motivating factors for entrepreneurs include autonomy, personal fulfilment, a strong network of contacts, a desire for riches, power, and an improved social position, as well as family support and cooperation. For women, the most important aspect in starting, running, and successfully completing a business venture is the sense of personal satisfaction that comes from overcoming obstacles. When a person's strong ambition to achieve isn't satisfied by a salaried job or when they want to turn a chance into a profitable idea, they should consider starting their own enterprise.

4.4.4 Policy Initiatives for Women Entrepreneurs in Bangladesh

The policy of the Bangladesh Bank to involve small businesses and a large number of women in financial operations has placed an emphasis on financial inclusion, according to the bank. Many policy measures aimed at empowering female entrepreneurs have been put in place by the Bangladesh Bank to accomplish this.

1. Funding options are provided at a discounted rate (bank rate + 4%), with banks and NBFIs obtaining refinancing at the bank rate, to help women entrepreneurs take on small- and medium-enterprise projects. An additional 15% of all refinancing opportunities are reserved for female business owners.
2. Women business owners can now access collateral-free loans of up to BDT 25 lakh through a newly instituted regulatory system that uses personal guarantees as security. The loan maximum has been set at 10,000 BDT, with group-based financing allowing up to 50,000 BDT, in an effort to increase the participation of marginalised and home-based women entrepreneurs. Cottage businesses now need a minimum credit score of 10,000 BDT, micro businesses, 20,000 BDT, and small businesses, 50,000 BDT.
3. The goal of this circular is to find three women in each branch's catchment area who are interested in starting their own business. As a means of enhancing their capabilities, these women will get specialised training in their chosen fields, and one of them will be granted access to financing.

4. Banks and non-bank financial institutions have been instructed to prioritise women entrepreneurs in the provision of loan facilities and to implement capacity-building activities for their current and potential women clientele. To facilitate efficient services for women entrepreneurs, it is recommended that banks and non-bank financial institutions establish dedicated "Women Entrepreneur Desks" in every branch. The desk is recommended for assisting potential women clients with the production of loanable company proposals, marketing plans, and related matters.
5. To enhance the oversight of women's entrepreneurial development programmes, banks and NBFIs are recommended to establish a WEDU at their respective head offices.
6. Bangladesh Bank has just launched a Simplified Loan Application Form in Bengali for SME entrepreneurs, especially women entrepreneurs. This program will alleviate challenges in fulfilling paperwork requirements.
7. Bangladesh Bank has released a circular granting a 3-month grace period for 1-year loans and a 3- to 6-month grace period for medium- and long-term loans for cottage, micro, small, and medium enterprises (CMSMEs).

4.4.5 Indian Government Programmes and Policies to Support Women Entrepreneurs

An essential driver of economic progress, women's entrepreneurship has been acknowledged. Entrepreneurial women not only generate new employment opportunities but also offer society novel approaches to long-standing issues in management, organisation, and business. But even so, they are just a small fraction of the total entrepreneurial population. Along with discriminatory cultural norms and laws around property, marriage, and inheritance; a dearth of formal financing methods; and restricted mobility, women business owners may face gender-based barriers, such as limited access to information and networks.

One way in which women can help achieve the Millennium Development Goals (MDGs) is by starting their own businesses. This can have a significant impact on family and community income, poverty levels, and women's empowerment. Consequently, there are several programmes, incentives, and promotional initiatives aimed at supporting women entrepreneurs, spearheaded by governments and developing organisations worldwide. Of all the small-scale industrial entities in India that are run by women, more than half are located in the four southern states plus Maharashtra (Subbarayudu & Rao, 2021).

To help women in India who are interested in starting their own businesses but might lack the necessary education or experience, several institutions are offering programmes such as the Entrepreneurship Development Programmes (EDPs). These include nationalised banks, MSME development organisations, state small industries development corporations, and non-governmental organisations. In addition, the Office of the DC (MSME) has established a Women Cell to aid women entrepreneurs by coordinating efforts and responding to their unique needs. Needy women can become economically independent through various government schemes, both federal and state, that offer financial aid to help with start-up training-cum-income-generating businesses. Additional programmes tailored to women business owners have been introduced by the Small Industries Development Bank of India (SIDBI).

In addition to dedicated programmes, numerous government initiatives aimed at micro, small, and medium-sized enterprises (MSMEs) offer unique advantages and subsidies to women entrepreneurs. For example, women are given priority under the Prime Minister's Rozgar Yojana (PMRY). In addition, the government has loosened up a few restrictions for women so that they can participate more easily in this programme. Likewise, MSMEs contribute 30–80 per cent of the total project cost under the MSE Cluster Development Programme. Nevertheless, the Ministry of MSMEs can contribute up to 90 per cent of the project cost when it comes to clusters run by female entrepreneurs. Similarly, under the Credit Guarantee Fund Scheme for Micro and Small Enterprises, women-run or -managed MSEs can receive 80 per cent guarantees; however, the total guarantee coverage cannot exceed 75 per cent of the loans that are given. The following are just a handful of the many innovative initiatives aimed at empowering female entrepreneurs by different branches of government and their related groups.

Pradhan Mantri Rozgar Yojana (1993) is a great social and financial programme for women entrepreneurs, and it goes by another name: PMRY. With the help of women entrepreneurs and bright minds in the workplace, this plan aims to facilitate skill-based self-employment so that people can achieve financial independence. Cost, eligibility, and subsidy cap revisions led to the development of this programme, which serves both urban and rural regions. Borrowers are limited to a maximum of Rs. 12,500 in loan subsidies, which is capped at 15 per cent of the project cost.

Udyogini Scheme (1997–1998) provides women with the financial tools they need to become economically independent, which promotes self-sufficiency and helps women grow as individuals. This women's loan plan is a reliable source of capital that supports aspiring women entrepreneurs by offering loans at reasonable interest rates compared to those in the private sector. Only households with annual incomes below Rs. 40,000 are eligible. Loans of up to Rs. 1 lakh are highly encouraged, with a special emphasis on the service and commercial sectors.

The Annapurna Scheme (2000) is such that women in the food-catering sector who are just starting with their own small companies are eligible for this credit. These women entrepreneurs can put the credit towards necessary capital expenditures like setting up vehicles, purchasing equipment and utensils, etc. One of the most popular ventures that women entrepreneurs pursue and succeed at is selling pre-packaged food items and snacks. This is because many women have grown up in the home and are comfortable with this type of work. Sales go up because they have a better shot at getting better capital and new items to launch their firm than they would have gotten otherwise. Within the plan, the maximum loan amount is Rs. 50,000.

Under Stree Shakti Yojana (2000–2001), women business owners can take advantage of a 0.05 per cent rebate on loans over 2 lakh and other perks. Any driven women entrepreneurs looking to get their businesses off the ground are welcome to apply. These women business owners must enrol in the Entrepreneurship Development Program (EDP) as a necessary condition.

The Synd Mahila Shakti Scheme (2001) initiative was started by the Syndicate Bank to empower women entrepreneurs. As part of this program, the bank is giving back to women company owners and self-employed people. Encouraging women to seek business loans to expand their operations is the main goal of this plan, which aims to empower them economically. The government's goal is to create an environment where women can easily get loans of up to 5 lakh so that they can expand their companies.

The Cent Kalyani Scheme (2006) is a particular focus on this government programme for women entrepreneurs, but it is open to new firms as well as those looking to expand. Neither the retail trade nor the educational or self-help sectors are qualified to participate in the programme. You may find the regulations and a list of qualifying categories on their website. In addition to not charging any processing costs, this loan does not require collateral. Through these women's loan programs, women entrepreneurs from certain qualified backgrounds can get the capital they need to grow their businesses. The maximum loan amount under this arrangement is Rs. 1 hundred million.

Bharatiya Mahila Bank Business Loan (2013) was first established to provide a large-scale support system for women entrepreneurs and their enterprises. It is considered one of the best government schemes for women entrepreneurs. A lot of women are trying their hand at entrepreneurship, and they're always hoping for the best. The goal of this bank is to help women achieve economic independence. Up to Rs. 20 crore can be borrowed from this source.

Mudra Yojana Scheme (2015) empowers women through access to business loans and other forms of financial assistance; this programme is part of the Indian government's effort to raise women's economic standing. They will be issued Mudra cards, which function similarly to credit cards, upon loan approval; the maximum amount that can be withdrawn is 10 per cent of the loan. Different sorts of programs are available under this scheme based on the type of business, the level of expansion, and the loan purpose. The government's women's loan initiative has a maximum loan amount of Rs. 10 lakh.

Mahila Udyam Nidhi Scheme (2015–2016) is a women's lending programme we have created to close the equality gap. For various industries, it encourages MSMEs and small-scale investments to thrive. As a result, there will be more motivation to restore SSI units that were previously written off as hopeless but that could be salvaged. The maximum amount that can be borrowed is Rs. 10 lakh, and the borrower is given 10 years to pay it back.

With regard to the Orient Mahila Vikas Yojana Scheme (2020), individual women or women acting as a proprietary concern with a combined 51 per cent ownership stake are eligible to apply for this women's loan programme. This is a fantastic chance for these stakeholders to grow their company and contribute to the advancement of their industry. These loans are available to women entrepreneurs in India at a concessionary interest rate of up to 2 per cent and do not require collateral. The maximum loan amount is Rs. 25 lakh, and the payback duration is adjustable, lasting up to 7 years.

With regard to the Dena Shakti Scheme (2020), farmers, retailers, manufacturers, small company owners, and micro-credit organisations are the only ones eligible for this government programme supporting women-owned businesses. Women beneficiaries are subject to maximum ceiling limits set by the Reserve Bank of India (RBI) that vary by the industry in which they intend to launch or grow their businesses. You can borrow up to Rs. 20 million.

When it comes to economic growth, women are the ones that shine. The situation is different in India, where some superstitions and restrictions on women are still practised. Just because a woman can run a household well doesn't mean she can't run a business just as well. Women have achieved success in every field imaginable, including teaching, law, engineering, medicine, economics, technology, and more.

The social, cultural, demographic, and geographical factors that influence women's success vary greatly from one nation to another. In addition to financial and governmental assistance, women needed other distinct motivating elements. To succeed as a woman entrepreneur, you need certain specific motivating elements. To be a successful woman entrepreneur, one must first understand the specific challenges that women entrepreneurs face.

Every country needs entrepreneurial women. To be competitive with advanced nations, we need gender parity in all spheres of society. Although men tend to do better, women should also experience good growth in their entrepreneurial endeavours. The government should institute programmes that help women advance in their entrepreneurial endeavours. If we are serious about helping women business owners succeed, we must do more than just propose programmes; we must methodically implement them. Women should have an easier time advancing in the entrepreneurial world due to specific cultural and socioeconomic variables. In addition to financial aid and government programmes, women need psychological support from their husbands and families if they want to succeed as entrepreneurs. Women may not be able to succeed as entrepreneurs, regardless of government assistance in financial matters, if they do not have the moral support of those closest to them.

4.5 CONCLUSION

Modernity is defined by lightning-fast progress in fields such as education, rural development, industrialisation, transformation, and technology. When this occurs, women's employment options grow substantially. Women fearlessly pursue entrepreneurial endeavours that benefit their families and the country as a whole. It is commonly believed that women in India, a predominantly male-dominated society, are reliant on men for both social and economic support. Family, the state, society, male co-workers, etc. all have a role to play in supporting women. They can join the national rural economy's mainstream and help advance rural economic growth with the correct support from the aforementioned groups. Indian women business owners could benefit greatly from government initiatives that aim to empower them via training, development, and capacity-building. Without proper training and financial backing, women business owners cannot grow. So, to help women succeed as entrepreneurs, we need to offer them the tools they need, like education and funding. Women in both urban and rural areas should have access to comprehensive educational programmes created by the government. To change the way society views women, the government should launch awareness and education campaigns. It is important to take into account the socioeconomic background of women entrepreneurs while designing training programmes. To help current and future rural women business owners overcome mental health issues, low self-esteem, and a lack of willingness to take risks, non-governmental organisations (NGOs), psychiatrists, technical experts, and management specialists should offer therapy. The challenges faced by women entrepreneurs cannot be resolved through governmental action alone. There need to be more women entrepreneurs who can think outside the box, build successful businesses, and persuade others to back their proposals. To overcome shared obstacles as entrepreneurs, we need to band together in self-helping, mutually beneficial organisations.

REFERENCES

Goyal, M., & Parkash, J. (2011). *Women entrepreneurship in India-Problems and prospects.* Zenith International Journal of Multidisciplinary Research, 1(5), 195-207.

Pallavi, D. R., Ramachandran, M., & Sivaji, C. (2022). *A review on entrepreneurship and its implication.* Trends in Banking, Accounting, and Business, 1(1), 36–42. https://doi.org/10.46632/tbab/1/1/8

Reddy, Y. S. (2024). The role of NGOs in promoting women entrepreneurship in India. *International Journal of Advanced Research in Science, Communication and Technology*, 6(1), 368–372. https://doi.org/10.48175/IJARSCT-19559

Saraswat, R., & Lathabahvan, R. (n.d.). *A study on women entrepreneurship in India.* Mukt Shabd Journal, 9(6), 3422–3432

Sharma, Y. (2013). *Women entrepreneur in India.* IOSR Journal of Business and Management, 15(3), 09–14. https://doi.org/10.9790/487X-1530914

Subbarayudu, M. C., & Rao, C. S. (2021). *Policies and schemes for women entrepreneurs in India.* International Journal of Research and Analytical Reviews, 8(1), 213–225.

5 The Dark Kak Rises

The Journey of a Small Startup Aiming to Redefine Reality

Zarjina Tarana Khalil and Tilka Farzana

5.1 THE STARTUP DREAM BEING DELIVERED

Bangladesh is experiencing a dynamic shift in its economic landscape, largely driven by an increasingly vibrant startup ecosystem. After a period of sluggish economic performance, the country is now poised for a significant economic upturn. This resurgence is particularly noteworthy given the challenging global economic climate. *The Economist* has recognized Bangladesh as "South Asia's standout economic performer" for 2023, underscoring its remarkable economic strides despite global uncertainties (The Economist, 2023).

A pivotal factor behind this resurgence is the country's burgeoning startup ecosystem. The ecosystem started its journey in early 2010, and multiple sectors show tremendous promise, including fin-tech, ed-tech, health-tech, and agri-tech (Saif, 2021). The country has made substantial progress in this domain, ranking 89th globally in the startup ecosystem and 114th in the international innovation index according to Startuprise (2024). This progress reflects a rapidly evolving landscape where innovation and entrepreneurship are becoming central to the nation's economic strategy, particularly due to the successful implementation of many national incubator and accelerator programs (Saif, 2021).

In the fiscal year 2023–2024, Bangladesh attracted an impressive $72 million in investments through 45 deals (Startuprise, 2024). This influx of capital highlights investor confidence in the potential of Bangladeshi startups, even though the sector faces significant challenges, including a relatively low rate of one in ten startups becoming successful. Nevertheless, the sheer volume of new ventures—over 2,500 active startups with more than 200 new ones emerging annually—demonstrates the robust entrepreneurial spirit driving the economy (Zavalko, 2023).

A youthful and tech-savvy population is one of the key reasons for the startup boom. With 65% of the population being under the age of 25, there is a talent pool eager to innovate and drive technological advancements (The Daily Star, 2023). Additionally, the country boasts a 75% internet penetration rate, which has significantly contributed to the growth and success of startups. The widespread adoption of digital technologies and online platforms has created a fertile environment for new business models and innovations.

DOI: 10.1201/9781003544838-5

The Bangladeshi government has also played a crucial role in fostering this growth by implementing supportive policies and frameworks aimed at nurturing innovation and entrepreneurship. Government efforts to streamline regulatory processes, provide funding opportunities, and facilitate access to resources are making it easier for startups to launch and scale. Notable success stories from the Bangladeshi startup scene include bKash, Pathao, and Chaldal. These companies have not only achieved significant milestones but also disrupted traditional markets with their innovative approaches. For instance, bKash has revolutionized financial transactions in Bangladesh by providing a mobile-based payment solution that is accessible to millions of people. Similarly, Pathao has transformed the way people commute and order goods with its ride-sharing and logistics services. The year 2023 has been particularly noteworthy for startups as they demonstrated significant resilience and dynamism amidst global economic uncertainties (Sakib, 2024).

Bangladesh's youthful population is marked by its adaptability, creativity, and receptiveness to new ideas. Startups are leveraging this vibrant energy to tackle local issues while seizing global opportunities. As the country continues to invest in and nurture its startup sector, it is well-positioned to sustain its impressive economic growth and further enhance its standing as a dynamic player in the global innovation landscape. The enthusiasm and participation of the Bangladeshi people are essential to this journey, as they remain deeply engaged with and supportive of the nation's entrepreneurial ambitions.

5.2 WOMEN IN STARTUPS

Women are underrepresented in the startup sphere globally, and hence, the startup culture in Silicon Valley is often referred to as one of "tech bros" (Kovaleva et al., 2022). Even in a country like Finland, which ranks very high in terms of gender equality, there exist gender-based barriers, including cultural stereotypes and lack of role models (Kovaleva et al., 2022). Surmounting the constraints of a traditionally patriarchal society, women in Bangladesh are increasingly making their mark in the startup sector, carving out a significant presence and reshaping the entrepreneurial landscape. This emerging wave of female entrepreneurs are not merely participating but driving innovation and growth across various industries, from e-commerce to healthcare and education.

Women entrepreneurs are harnessing their entrepreneurial spirit and innovative prowess to launch and scale ventures that are transforming their respective fields. For instance, Roksana Parvin Tuhin's Chhaya Mohila Unnayan Sangstha is a notable example of how women-led initiatives are addressing social issues and empowering local communities. Similarly, Monira Munni's efforts to preserve cultural heritage through hand-stitched products not only showcase traditional craftsmanship but also tap into niche markets abroad, demonstrating the global appeal and marketability of Bangladeshi artisanal work (The Business Post, 2021).

The startup ecosystem in Bangladesh is further enriched by tech-driven ventures like SBK Tech Ventures, which plays a pivotal role in investing in and nurturing women-led startups. By providing crucial financial support and resources, SBK Tech

Ventures helps these startups overcome initial barriers and scale their operations (Dhaka Tribune, 2024). Platforms like FutureStartup are instrumental in highlighting the achievements of female founders, offering them visibility and ensuring they serve as a source of inspiration for the next generation of women entrepreneurs. However, it is important to note that, even with the available opportunities, women's participation in entrepreneurship is on the lower end both for developed countries as well as developing countries (Kuschel & Lepeley, 2016).

Despite the progress, there are considerable challenges thwarting women-owned ventures from reaching their potential. They face significant hurdles, such as limited access to funding—evidenced by the fact that 93% of all loans given to startups are directed toward male-led ventures (Parveen, 2024). Gender bias remains a barrier, influencing perceptions and opportunities in the startup ecosystem (Saifuddin et al., 2019). Additionally, in low-income countries (such as Bangladesh), women have few chances to find jobs and even fewer chances to start a business, so they often end up participating in a very basic form of entrepreneurship, restricted mainly to buying and selling products, rather than finding new solutions to existing problems (Kuschel & Lepeley, 2016). They also do not have much scope or access to decision-making roles (Saifuddin et al. 2019). Nevertheless, the growing presence of supportive infrastructure, including incubators, accelerators, and government-backed initiatives like Startup Bangladesh, provides a foundation for these entrepreneurs to thrive (Monamee, 2024).

Yet, the path forward is not without obstacles. Regulatory barriers and limited mentorship opportunities continue to pose challenges that need addressing so as to sustain and amplify the momentum of women's participation in startups. Ensuring equal access to resources and support will be crucial in unlocking the full potential of female entrepreneurs. By tackling these issues, Bangladesh can fully leverage the talents and innovations of its women, driving further economic growth and achieving remarkable success in the global startup arena.

While the current landscape is marked by significant achievements and promising developments, it is essential to continue addressing existing barriers and providing robust support to female entrepreneurs. Due to the lack of gender profiling of entrepreneurs as individuals, there is a need to share the unique narratives of female entrepreneurs (Williams et al., 2020), and the case study of this chapter is an initiative toward that.

5.3 WOMEN IN TECHNOLOGY

The gender imbalance in technology and the sciences is a gaping truth in the global sphere, and even recent statistics validate this imbalance (Kovaleva et al., 2022). The remarkable advancement of women in technology in Bangladesh is a testament to their resilience and the support provided by various initiatives. Despite encountering significant challenges such as gender bias and limited access to education, women in Bangladesh have made notable advances in the tech and digital education sectors, driven in large part by the country's impressive internet penetration rates. Additionally, the exposure to computers and the internet from an early age is positively influencing girls to pursue tech-based majors and careers (Saifuddin et al. 2019).

The resilience of these women is complemented by the efforts of organizations and initiatives dedicated to fostering their growth in technology. Organizations such as Women in Digital and initiatives like the Bangladesh Women in Technology (BWIT) are crucial in providing training, mentorship, and networking opportunities, fostering a conducive environment for women in tech (Kabir, 2023). They provide platforms for women to learn and grow, helping them convert their ideas into reality.

Government policies also play a crucial role in promoting digital literacy and inclusion. By focusing on digital skills development, the government can ensure that women have the tools and knowledge needed to excel in the digital economy. The proliferation of digital platforms and e-commerce has further opened up new opportunities for women. These platforms not only provide new business avenues but also enable women to reach broader markets and scale their ventures effectively. The accessibility of digital tools and resources has been instrumental in helping women overcome traditional barriers and contribute meaningfully to the tech industry, and there is plenty of scope for far-sighted entrepreneurs since tech-based businesses are one of the fastest-growing sectors (Kuschel & Lepeley, 2016).

The need for more visible role models remains a significant challenge. Women in tech can benefit greatly from seeing more examples of successful female leaders and innovators. Additionally, solutions for better work-life balance are essential, as many women struggle to manage the demands of their professional and personal lives. Workplace discrimination also continues to be an issue that needs addressing so as to create a more equitable and inclusive environment.

Despite these challenges, the increasing presence of women in technology in Bangladesh is a positive indicator of the country's evolving tech landscape. The growing number of women entering and excelling in tech roles are not only enhancing innovation but also contributing to broader economic growth. As the sector continues to develop, the continued support for women in technology will be vital to sustaining this momentum and ensuring that the benefits of digital advancement are shared equally.

Kovaleva et al. (2022) developed a model summarizing the factors that influence the intention of females to take up entrepreneurship and the consequences of those factors. The case below, Dark Kak, will be used to illustrate how these factors impacted the venture and the actions taken by the founders to survive, as shown in Figure 5.1.

The factors include:

1. Social factors
    a. Lack of female entrepreneurs.
    b. Pressures of stereotypes in childhood.
    c. Pressure of being family caretakers.
2. Personal factors
    a. Empathy.

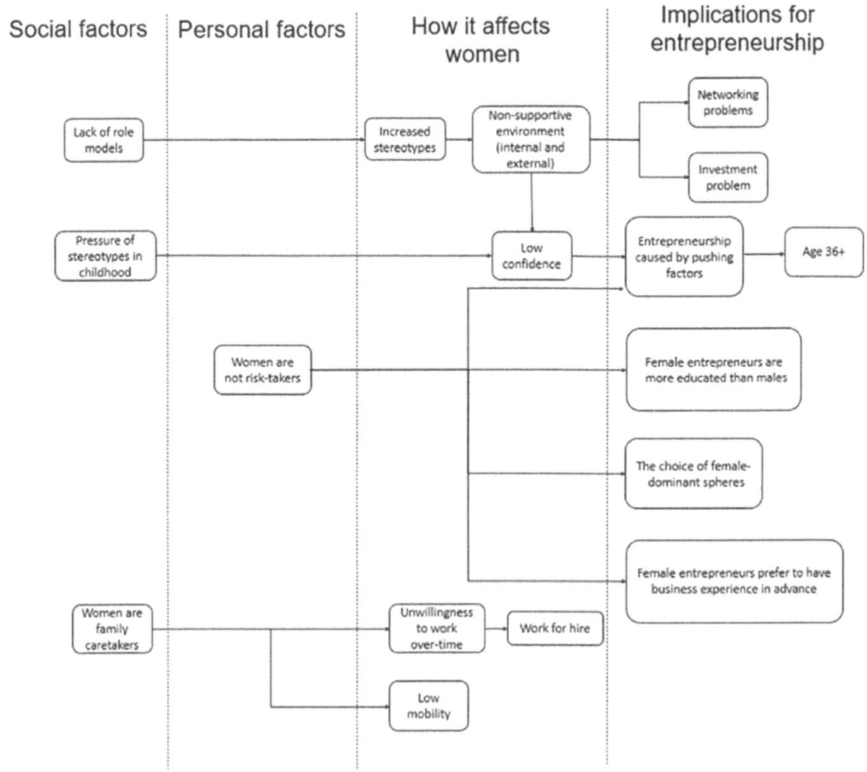

FIGURE 5.1 Factors Affecting Women Entrepreneurs.

   b. Family of entrepreneurs.
   c. Essential knowledge.
   d. Rich field experience.
   e. Perfectionism.
3. Impact on women
   a. Focus on social impact (due to empathy).
   b. Leadership potential (family support).
   c. Self-confidence (family support).
   d. Government support (required to mitigate the lacking).
   e. Self-doubt (due to stereotypes and perfectionism).
4. Implications for entrepreneurship
   a. Internal motivation (passion/empathy/leadership).
   b. Business ideas (due to knowledge and background).
   c. Late entry into entrepreneurship.
   d. Additional attention for being a female entrepreneur.
   e. Quotas and networking activities (due to government support).
   f. Investment problems (due to stereotypes).

5.4 DARK KAK

Dark Kak is a tech company that mixes augmented reality/virtual reality (AR/VR) and Internet of Things (IoT) to make immersive experiences that change how we see things. It started as a personal venture of the founder Nourin Haque Ridi, an engineer by profession. Just out of personal interest, she would create animated AR versions of photos, where pictures come to life, and share them on social media. She started to notice an active interest among her contacts. People started reaching out to her, asking her to animate their photos of their loved ones, to be kept as memorabilia. The photos could be framed with a QR code, and through that, those special videos could be played. This concept became popular, and she started to receive more and more of these orders. However, although this buzz and excitement were garnering interest, Ridi knew this was just some additional income alongside her regular job. Unless she really pushed the boundaries, it would not be a "business". Hence, she thought it was now time to think in terms of a bigger picture.

The name Dark Kak stemmed from the fact that since the founders were all jobholders, they would usually work together in the evenings and late into the night—hence the word Dark. "Kak," meaning crow in Bangla, symbolizes intelligence and inventive grit—figures often celebrated in folklore for learning through trial and reinvention. The name also nods to the team's favorite film, *The Dark Knight Rises*.

5.5 SOCIAL FACTORS

When studying in universities, students do not typically think about startups. Especially in government universities in Bangladesh, the culture or concept of startups do not exist. However, the trend is changing, and students are now more willing to work in collaboration and not necessarily in competition. This is a positive change and will make the operation of startups more conducive. The game industry community is very small, and the number of technical people is very limited. Hence, there is a lot of dependence on online resources. Lots of people may have nice ideas regarding the usage of AR/VR, but they do not have the technical knowledge. Typically, women are less involved in this sector, and when it comes to recruitment, particularly to higher positions, there is an unspecified bias (it is perceived that women will not be suitable for the job). Female game and software developers are very rare. Furthermore, with the use of AR and VR, it is easy to create hype but difficult to hold on to it; hence, slower performance due to bandwidth limitations and device issues often proves to be detrimental. Building awareness and directing people with the right set of instructions remain an ongoing challenge.

For the founder Nourin, the journey has been tumultuous, to say the least. Coming from a professional family, her foray into entrepreneurship wasn't exactly welcomed—not because they were unsupportive, but because the timing and circumstances didn't seem right. The idea of leaving a stable job to pursue something that entailed a lot of uncertainty was baffling for the family.

However, with time, she noticed that her family stood up for her in times of distress, which also made her realize that she would have to take a hard path and hold on to the needed stability of a job while building her venture, not just for herself but to keep things steady for her family, and her team as well. Hence, there was the pressure of added responsibility to her family.

5.6 PERSONAL FACTORS

A successful startup starts off with the passion of its founder. In addition to that, they need to have the perseverance, the quest for learning, and the openness of mind to imbibe new and innovative ideas and business practices (Workplace One Business Centres Limited, 2025). Personally, the ability to deal with failures and resilience are two primary characteristics which Nourin feels she has and which are necessary for startup founders. In her initial days, at times she struggled to explain to people the nature of her work, so much so that, for about a year, she would just tell people that she was unemployed to avoid the long conversation.

Additionally, she considers herself to be an observer, particularly observing people in leadership roles, and tries to imbibe characteristics which she deems necessary. Another factor which is very important for startups is to be customer-centric. Since Dark Kak is a small and young startup with a limited number of projects, at this point, they are able to listen to each of their clients and provide customized packages as per their requirements. Of course, going forward, once they become bigger, they would still want to prioritize this close relationship with clients as she would like to build a customer-centric culture. She also feels that her team operates in the same manner, whereby they are continuously trying to improve themselves and learn from each other and lift each other up. At times, when working with third parties, if they observe any particular input or practice which is admirable or useful, they try to imbibe that.

At present, Ridi has to balance a full-time job with the venture, ensuring security for her parents. Although that takes a toll on her work-life balance, the trade-off is that she gets to work on her dream. And that dream? It's about supporting others, helping them work toward their own ambitions. She looks forward to the day when Dark Kak can grow to employ and inspire hundreds of other dreamers, alongside her co-founders, and to the day when their parents would be proud.

5.7 IMPACT ON WOMEN (SOCIAL IMPACT, LEADERSHIP POTENTIAL, AND CONFIDENCE)

In her early days, Nourin would visit anime expos with animated cards. Through her interactions and observations, she noticed that ordinary people required a great deal of customized instructions to experience tech-based animations. Additionally, when customers bought an item but could not operate it at home, they would complain that things were not working properly. One of the major causes of this is internet bandwidth and Android/iOS versions; and in the AR and VR domain, if things do not work due to bandwidth, people assume that it's a technical problem. Hence, it became important to raise awareness regarding these issues. Considering the fact that bandwidth is a national challenge, she realized that the B2C market (individual customers) would be very difficult for her as it was not going to be profitable or sustainable. In the meantime, two of her partners joined Dark Kak; Shababa Binte Hossain was an EEE major and Md. Nazmul Islam Nishad was a CSE graduate. They were also interested in augmented reality and wanted to work with IoT. This added to their expertise, and with time and experience, they aimed at making more relevant items.

Although Dark Kak started as a personal endeavor where the founder would animate photos, she realized the power of visuals and the impact it had on the human mind. While participating in many hackathons, the whole process of game development and then converting that to augmented reality became very addictive for her. However, there were two particular events that further motivated her to stay on. Firstly, they created a project called "Kind New World" where, on one particular day, children would have to fulfill a challenge like "make a new friend" or "help a stranger". This game was used by a group of volunteer teachers from the Platform of Community Action and Architecture (POCAA) in a school in the slums of Gabtoli (a crowded suburb outside Dhaka) where children could make and share videos of a "kind activity". Similarly, a school in remote Bandarban named Pawmang was able to raise around TK 25,000 by using a VR art exhibition (paintings of students from the school were photographed and exhibited virtually). These kinds of social impact motivated the founders to continue in this domain, as it gave them confidence that they could make a decision. More is shown in Figure 5.2.

The benefits of technology are both tangible and intangible as they empower women to make money, drive results, and meet customer demand (Saif, 2021). Dark Kak received an award from She Loves Tech for a project for smart jewelry. Their target was to create a piece of jewelry that would also be a security device for women. At present, they are working on this in the R&D phase as they wish to design a

FIGURE 5.2 Kind New World.

local solution (applicable for Bangladesh), but it is a challenging initiative due to the hardware requirements. The aim is to make it user-friendly as well as cost-effective. Additionally, they wish to be trailblazers as females in the field of AR/VR so that more women are encouraged to take up projects in this domain, in gaming, software, and app development.

5.8 IMPACT ON ENTREPRENEURSHIP

The augmented reality game market is a booming market in the country, and recently, another local company, Studio 23, has been invited to the Gaming Industry Conference (GIC) in Poland (The Daily Star, 2024). The interactions with the market made Dark Kak identify a gap, and they came up with the AR mirror, which served as an affordable solution as VR headsets were expensive whereas individual games for B2C came with addressing technical issues. The AR mirror enabled people to interact with props and products in Kiosk and any smart board or any display with an Android or Windows OS, as shown in Figure 5.3. Most retail spaces nowadays have kiosks and screens which constantly play ads. They wanted to use that spot to create an interactive space where customers could pose, use hand gestures, and take pictures with products, and they finally pitched the idea to Grameen Danone through another, third-party, company. Luckily for them, Grameen Danone loved the idea and was ready to bring them on board and introduce them to a kids' fair.

FIGURE 5.3 AR Mirror for Children.

At the kids' fair, Dark Kak kept tabs on where people could play games, and they also leveraged the kiosks where people could interact with the products. Although they were supposed to provide technical facilities at the kids' fair, they ended up having such a high level of engagement that it opened up another avenue for them, particularly with regards to engagement with children. They realized that even months after the event was over, people were still playing the games online and not just people from Bangladesh but also from other countries like India, Thailand, and Indonesia. Although the activities were mainly on the kiosks during the event, the same games were deployed on their website and social media, and the engagement continued.

At this point, they thought of leveraging the AR mirror in the fashion industry. Hence, for Fashion Tech Week 2024, organized by Apparel Resources from India, they designed a mirror where people could virtually see how they looked with a product on, and also, they allowed virtual store visits through VR headsets. The third concept, which they tried for fun, was an augmented t-shirt that meant that once you wore the t-shirt, the design would pop out in an animated manner. All three concepts were met with massive enthusiasm, which was beyond their expectations.

This tech week proved to be a game changer for Dark Kak as it brought them into the spotlight. Soon after this event, the approach of their clients toward them had changed. Since Bangladesh has a solid ready-made garments (RMG) sector and a fashion sector, this industry started to show much interest toward the use of these technologies in their own companies. Dark Kak, at this point, tries to offer them customized packages based on the clients' stores as well as interactive event attractions. The constant trials and errors were driven by their continuous pursuit of perfection and the intention to provide the customers with a seamless experience.

5.9 BUSINESS MODEL

At present, Dark Kak operates on a package basis. Companies approach them and lay out their requirements; based on that, Dark Kak tries to customize a package. Since the industry is still new, customers may sometimes have some unique demands, and a customized package is offered based on the requirements.. For augmented reality, a standard package usually includes interactive gaming experiences—delivered through tools like the AR Mirror or the newer Magic Window, an AR display that functions like a virtual portal at the user's location, along with other customized interaction types. For virtual reality (VR), they are able to design virtual shops and e-commerce sites. The number of elements required depends on clients' needs, and additional elements are subject to extra charges. In terms of capacity, at this moment, they can manage three to four projects. However, during the event seasons, there is a high rush, and it becomes necessary to get some work done through third parties.

5.10 PROSPECTS IN BANGLADESH

The prospect of this industry is very high but there is a lack of awareness in the country. If there were a higher level of awareness, there would definitely be a huge market in the marketing and branding sector. The fashion industry has recently imbibed parts of the practice, but there is huge potential in the education and healthcare sector, although the industry needs to develop its own capabilities for making global-standard products. Alongside, if the bandwidth issue can be addressed, there will be faster and easier adoption of these products. One of the ways Dark Kak is trying to mitigate the issue is by developing products that are able to operate offline, without the support of the internet. Additionally, being tech-people themselves, the founders are continuously trying to optimize their software (e.g. reduced loading time) to give the users a smooth and seamless experience, since they know that today's customer also has a short attention span. Hence, the improvisations are based on the local market, and eventually, they would like to make a product of global standards so that they can operate on a global subscription model.

5.11 HOW BRANDS CAN LEVERAGE THIS SERVICE

After the fashion tech week, they got the opportunity to understand the RMG sector. The majority of RMG companies have expatriate buyers who often would like a virtual tour of the factories. Usually, in absence of a virtual tour, photographs of the factory are shared. However, given the increased focus on compliance and alignment with SDG, a virtual tour better showcases the initiatives of the factory, and there is a huge scope in this sector. Additionally, virtual training for the operations of machinery is another very lucrative segment as it speeds up the process and makes workers more confident and adept at operations. When it comes to the retail stores, the AR trial includes the ability to see sizes and colors online. It engages people by giving them the ability to try items from a distance and also take fun selfies.

For non-fashion items, using interactive packaging also engages customers—for example, in their work with Grameen Danone, they focused on interactive packaging of Shokti Doi in which, through their animations, they tried to highlight both the taste and nutritional value of the product. This also helps in generating subliminal perceptions, and these can be used on a range of products, like highlighting eco-friendliness or reducing the use of plastic.

5.12 CHALLENGES AND THE WAY FORWARD

The startup community is booming, and there are several organizations that are working toward grooming and nurturing the startups. Government-backed initiatives have a bigger community, but female founders are limited there. The number of female startups working in partnerships is very limited, and of course, it should grow in the coming days.

One of the main aspirations for Dark Kak at the moment is to create an active community where there will be more awareness regarding AR and VR and also more people interested to learn about them. Hence, they are planning to introduce an AR Walk, a concept similar to the Photo Walks organized in universities. Through this AR Walk, the purpose is to visualize how the images around the city may be animated or presented virtually. The idea is also to engage the community so that once demand is higher, it can be mitigated through the community members.

So far, they have received three awards, one from She Loves Tech and UNESCAP, one from Apparel Resources 40 under 40, and a "Pass It On" award from Anitab.org. The award from She Loves Tech and UNESCAP helped them gain attention, but what helped even more is the ongoing relationship which they hold with the organization. Hence, it didn't just end with one award; rather, they received continuous guidance and mentorship on a customized basis. Sharon Lim (chief content officer of She Loves Tech) visited Bangladesh, met Dark Kak, and provided valuable advice, and also, the whole community provided highly valuable networking opportunities. As a winner of the "Pass It On" Award from Anitab.org, the organizers arranged a meet and greet with the GHC'24 conference for them to network and share stories with the global community of women in tech and entrepreneurship. The award from Apparel Resources also put them under the spotlight in the fashion industry, through which they were able to showcase their key activities to a wide audience. The awards and recognition have helped the venture to gain momentum, but of course, more support groups like these would further enrich the bonding between startups and possibly encourage more women to join tech ventures. At present, Dark Kak is fully self-funded, but the recognition, mentorship, and networking opportunities have instilled confidence in the founders, and that would also help them to pitch for funding in the coming years.

REFERENCES

The Business Post. (2021, March 12). Women lead many successful startups in Bangladesh. *The Business Post*. https://businesspostbd.com/economy/women-lead-many-successful-startups-in-bangladesh-15024

The Daily Star. (2023, February 28). Digital Bangladesh: the story of transformation, resilience, and sustainability. *The Daily Star*. www.thedailystar.net/opinion/views/news/digital-bangladesh-story-transformation-resilience-and-sustainability-3222626

The Daily Star. (2024, October 2). Bangladeshi video game invited at Game Industry Conference. *The Daily Star*. www.thedailystar.net/tech-startup/news/bangladeshi-video-game-invited-game-industry-conference-3717921

Dhaka Tribune. (2024, May 17). *SBK Tech invests $7.1m in 6 Bangladeshi startups* [Press release]. Retrieved August 31, 2024, from www.dhakatribune.com/business/346566/sbk-tech-invests-7.1m-in-6-bangladeshi-startups

The Economist. (2023, March 2). Is Bangladesh's admired growth model coming unstuck? *The Economist*. www.economist.com/leaders/2023/03/02/is-bangladeshs-admired-growth-model-coming-unstuck

Kabir, S. (2023, March 8). Bangladeshi women in tech and digital education: challenges and strides. *The Daily Star*. www.thedailystar.net/supplements/international-womens-day-2023/news/bangladeshi-women-tech-and-digital-education-challenges-and-strides-3266051

Kovaleva, Y., Hyrynsalmi, S., Saltan, A., Happonen, A., & Kasurinen, J. (2022). Becoming an entrepreneur: a study of factors with women from the tech sector. *Information and Software Technology*, *155*, 107110. https://doi.org/10.1016/j.infsof.2022.107110

Kuschel, K., & Lepeley, M. T. (2016). Women start-ups in technology: literature review and research agenda to improve participation. *International Journal of Entrepreneurship and Small Business*, *27*(2/3), 333. https://doi.org/10.1504/ijesb.2016.073995

Monamee, M. I. (2024, March 8). Breaking barriers: unveiling the challenges faced by women entrepreneurs. *The Daily Star.* https://www.thedailystar.net/tech-startup/news/breaking-barriers-unveiling-the-challenges-faced-women-entrepreneurs-3562001

Saif, S.B. (2021). Technology based entrepreneurship and its prospect in Bangladesh. *International Journal of Trade and Commerce-IIARTC, 10*(1) 190–209.

Saifuddin, S., Dyke, L., & Hossain, M. S. (2019). Walls all around: barriers women professionals face in high-tech careers in Bangladesh. *Equality Diversity and Inclusion an International Journal, 38*(7), 705–726. https://doi.org/10.1108/edi-11-2017-0247

Sakib, A. (2024, February 17). Startups are fuelling the Bangladesh of the future. *The Daily Star.* www.thedailystar.net/anniversary-supplement-2024/innovate-educate-elevate/news/startups-are-fuelling-the-bangladesh-the-future-3545086

Startuprise. (2024, April 9). Rising startup ecosystem of Bangladesh – Startup Rise Asia. *My Blog.* https://startuprise.org/rising-startup-ecosystem-of-bangladesh/#google_vignette

Williams, S. D., Ammetller, G., Rodríguez-Ardura, I., & Li, X. (2020). Narratives of international women entrepreneurs: An exploratory case study of identity negotiation in technology startups. *IEEE Transactions on Professional Communication, 63*(1), 39–51. https://doi.org/10.1109/tpc.2019.2961016

Workplace One Business Centres Limited. (2025). 5 common characteristics of successful startups. https://workplaceone.com/blog/5-common-characteristics-successful-startups/

Zavalko, M. (2023, August 11). Webinar recording | The rise of Commonwealth entrepreneurs: trials and triumphs of the Bangladesh startup ecosystem. Commonwealth Chamber of Commerce. https://commonwealthchamber.com/webinar-recording-the-rise-of-commonwealth-entrepreneurs-trials-and-triumphs-of-the-bangladesh-startup-ecosystem/

6 Impediments for Women in the Tech Industry

A Bangladesh Perspective

Nabila Kamal, Effat Binte Kabir, and Rani Thanacoody

6.1 INTRODUCTION

The participation of women in the labour market has long been a focus of academic and policy-level discourse in Bangladesh (Hossain et al., 2021). Bangladesh has made major advancements in recent years towards women's empowerment, emphasising the transition from the old state of stereotyping to contemporary gender dynamics. Women in Bangladesh are now more empowered, independent, and growth-oriented (Tabassum & Nayak, 2021). Nonetheless, several social, economic, and infrastructural obstacles persist that hinder women's career advancement. The presence of glass ceilings in businesses and cultural preconceptions hinder women's advancement in their entrepreneurial ventures in various industries, and the tech sector is no exception (Basharat & Alam, 2024).

The economy of Bangladesh is among the fastest growing in the world, and it is emerging as a new Asian tiger (Alam et al., 2020; Islam, 2019a). Bangladesh's recent advancements in the technological landscape have been regarded as a key factor fuelling its remarkable economic expansion (Mujeri & Mujeri, 2020). Bangladesh's technological landscape has grown rapidly, with key advancements in sectors like mobile financial services (MFS), e-commerce, and digital infrastructure. The country is now eyeing further development in the digital landscape with stronger internet connectivity, automation, and artificial intelligence (Afroja & Islam, 2023). On the other hand, women's entrepreneurship has been recognised as pivotal for the economic sustainability of developing nations (Gerguri-Rashiti & Rotabi, 2021). Consequently, relevant governmental entities and policymakers in Bangladesh have advocated for the advancement of women-led entrepreneurial ventures (Aziz, 2020). Bangladesh has experienced a surge in the emergence of women-led tech startups such as Foodpanda, Shohoz, and Prava, which have been flourishing in recent times (The Business Standard, 2023). However, despite the tremendous technological advancement, there persists a gender disparity with regard to women's entrepreneurship in Bangladesh's tech industry. Women in the country are still underrepresented in many IT-related fields and are more likely to be passed over for new and emerging roles (Kabir, 2023).

In the patriarchal society of Bangladesh, traditionally, women have occupied an underprivileged position and have been stumbling across various societal, economic,

DOI: 10.1201/9781003544838-6

and infrastructural impediments that have circumscribed their participation in entrepreneurial ventures (Lata et al., 2021). Nonetheless, Bangladesh has made strides in expanding the number of women employed (Chowdhury, 2022). Over the past few decades, Bangladesh has also achieved amazing strides towards the empowerment of women (USAID, 2024). Women empowerment by ensuring increased participation of women in the labour force becomes even more crucial for Bangladesh as the country seeks to fulfil the Sustainable Development Goals of the United Nations (Hossain, 2021). Nevertheless, Bangladesh's progress in boosting women's employment is still sector- or industry-specific. For instance, substantial growth in the number of women working has been observed in the agriculture and service sectors of Bangladesh (Bangladesh Bureau of Statistics, 2018). It is also noteworthy that Bangladesh has demonstrated remarkable success in leveraging technology to strengthen women's participation in online entrepreneurial ventures (Hasan et al., 2024). The small and medium business sector, which contributes an estimated 25% of Bangladesh's GDP, is a prime example of this (Sultana, 2021). Nonetheless, as mentioned earlier, women's representation in the tech sector in Bangladesh remains noticeably low (Kabir, 2023). This is especially true for women-led tech startups in Bangladesh (International Trade Center, 2015). Such underrepresentation of women in the tech industry in Bangladesh is crucial to address given the country's aim to establish itself as an economically resilient nation and its dependence on technological advancement for its sustainable economic development.

6.2 TECHNOLOGICAL REVOLUTION IN BANGLADESH

Our societies have undergone significant transformations due to waves of technological revolutions during the past few decades (Rahman & Abedin, 2021). With the emergence of the Fourth Industrial Revolution (4IR), organisations around the globe are facing novel challenges in their attempt to strive sustainably (Islam et al., 2023). 4IR's goal is to change the ways that information services are disseminated and executed across the world (Rahman & Abedin, 2021). The convergence of technological advancements has blurred the edges between the physical, digital, and biological realms, characterising the Fourth Industrial Revolution (Ashaari et al., 2021).

Bangladesh has emerged as one of South Asia's leading economically progressive countries (Azran, 2024). A 6.4% average annual rise in real GDP was underpinned by a steady macroeconomic climate from 2010 to 2023. In 2020, due to the COVID-19 pandemic, the growth in GDP was interrupted. Nevertheless, Bangladesh showed resilience in its economic growth during the pandemic. Bangladesh is projected to attain middle income status by 2031 and high income status by 2041 (World Bank, 2024). It has been advocated that novel fourth-generation technology in the industrial sector be used to enhance resilience, effectiveness, and economically sustainable manufacturing procedures in the future (Kamarthi & Li, 2020). Hence, Bangladesh has been focusing on sustaining its economic development and achieving the goal of becoming a middle-income country through sustainable technological development. The country's quick economic development has been further accelerated by the flourishing IT and telecom industries, as well as the rapid expansion of its youthful labour force (Alam & Ogawa, 2024). However, Bangladesh is still in the primary stage of the Fourth Industrial Revolution (Akter et al., 2024).

Nevertheless, within the dynamic global trade milieu, Bangladesh has demonstrated commendable progress in transitioning its business transactions to digital platforms (Afroja & Islam, 2023). Several ICT initiatives have been put into practice to promote digitisation, encompassing a range of sectors such as security, general services, utility, transportation, and educational services. Bangladesh's digitisation effort is mainly focused on four areas: improving ICT-based economic prospects, connecting citizens, building human resources, and delivering services to citizens' doorsteps (Islam & Inan, 2021).

In terms of connection and other ICT metrics, the last 10 years have seen incredible advancements. Between 2000 and 2010, the percentage of people using mobile phones increased significantly, reaching notable heights of 103% in 2020 (Hossain, 2020). There has been progress in improving internet connectivity, as the proportion of individuals with internet access has grown from 5% in 2012 to approximately 25% in 2020 (Hossain, 2020). Till August 2024, Bangladesh had reported 192.43 million mobile subscribers and 140.50 million internet users (Bangladesh Telecommunication Regulatory Commission, 2024a, b). As of 2022, Bangladesh had the 10th highest number of mobile subscribers in the world (CIA World Factbook, 2024). Despite being satisfactory, the progress is insufficient for having a major effect on the internet economy (Hossain, 2021).

Over the preceding two decades, Bangladesh has made noteworthy progress in fintech as well. Offering new avenues for conventional banking, fintech has already disrupted conventional financial services and has made financial services accessible to previously untapped customers (Basole & Patel, 2018). With rapidly improving technology, Bangladesh's successful fintech ecosystem has emerged especially with mobile financial services (MFSs) (Mahmud et al., 2023). According to 2021 data from the Bangladesh Investment Development Authority, 759 registered MFIs were serving 30 million clients with a credit portfolio of USD 10 billion and savings of USD 6 billion. Such data shows that Bangladesh has the potential for further development in various aspects of fintech. Despite these developments, economic and infrastructural barriers persist, especially for tech startups that often have limited access to financing.

6.3 OVERVIEW OF THE TECH INDUSTRY IN BANGLADESH

With technological progress in the country, Bangladesh's tech industry has also evolved. The industry is now thriving with potential and mainly comprises software development and IT-enabled services, including but not limited to business process outsourcing (BPO), enterprise resource planning solutions, and e-commerce. Such IT-enabled services have become one of the promising export-potential sectors for Bangladesh (Bangladesh High Tech Park Industry, 2024). For instance, in 2019, export earnings from the industry were recorded to be USD 1 billion. A sizeable pool of online workers further boosts their revenue. With 650,000 registered freelancers,

Bangladesh had the second-largest online labour pool globally in 2017, accounting for 16% of the market, second only to India (24%). Although software and application development products and services account for most of the industry's revenue, there has been a discernible rise in product offerings in areas such as data analytics, artificial intelligence, the Internet of Things (IoT), and augmented reality with e-commerce-based organisations thriving since the COVID-19 pandemic. According to the Bangladesh Investment Development Authority (n.d.) and industry reports by Upturn Bangladesh (n.d.), there are over 4,500 IT/ITES firms in Bangladesh employing more than 750,000 ICT professionals.

The tech sector stands as a formidable catalyst for job creation in Bangladesh's labour market. The sector is currently offering 300,000 jobs, which is expected to be 500,000 by the year 2025. It is also anticipated that this figure will be doubled by 2028 (Bangladesh Association of Software & Information Services, 2024). Bangladesh's government has been accentuating the integration of technology within the country's development initiatives. One of the notable initiatives taken by the government that had a significant positive impact is the Bangladesh Hi-Tech Park Authority. This was established to foster an atmosphere that is conducive to investment and creating jobs by stimulating the expansion of the country's tech industry (Bangladesh High Tech Park Authority, 2024).

The tech industry in Bangladesh has experienced rapid expansion largely due to advancements in internet access, mobile connection, and increased investments, making it a major force in Bangladesh's economy. In addition to providing new job possibilities, this industry, which includes software development, IT services, telecommunications, and digital platforms, among other things, helps diversify the nation's economy (Barry, 2018). Below, we discuss the major sectors within the tech industry in Bangladesh.

6.3.1 IT Outsourcing and Software Development

IT outsourcing is booming in Bangladesh for call centre operations, rule-based data processing, image processing, graphic or web designing, and basic data processing (Bangladesh Investment Development Authority, 2019). Nevertheless, it is worth mentioning that nearly two decades ago, the IT outsourcing and software development industry in Bangladesh was mostly a hardware vendor market with little or no local value addition. Bangladesh has seen the largest year-over-year growth in developer accounts, making it a global leader in software development growth, according to new data from GitHub, a well-known platform (The Daily Star, 2024).

6.3.2 Digital Platforms

The prevalence of digital platforms in Bangladesh has increased due to enhanced internet access and a greater abundance of mobile devices. Particularly in urban areas of the country, food delivery services and ridesharing applications have become widespread (Khatun et al., 2021).

6.3.3 Telecommunications and Networking Sector

The telecommunication sector in Bangladesh comprises mobile or GSM (Global System for Mobile Communications) operators, handset manufacturers, infrastructure, distributors, and retail content. At present, there are four GSM operators in Bangladesh. These are significantly contributing to the growth of the internet subscribers of Bangladesh as well (Bangladesh Investment Development Authority, 2021). The telecommunication sector continues to experience growth with increasing mobile and internet subscribers. According to a USAID study in 2019, the overall telecommunication sector in Bangladesh employed about 7.60 lakh people directly. However, it is noteworthy that only 7% of the employed workforce were women (Islam, 2019b).

6.3.4 Fintech and Mobile Financial Services (MFSs)

Mobile financial services (MFSs) are one of the most impactful developments in Bangladesh's tech industry, providing financial inclusion to the country's unbanked population. Mobile money has been used since the early 2010s for remittance transfers to families, and the COVID-19 pandemic and the growth of the e-commerce sector have made it increasingly simple for Bangladeshis to pay for services online (Massally et al., 2022).

6.4 WOMEN IN THE BANGLADESHI LABOUR MARKET AND WOMEN ENTREPRENEURSHIP

Over the past two decades, Bangladesh has made remarkable strides towards improving the lives of women. The country has been focusing on enhancing women's health, education opportunities, and employment. Bangladesh has been globally recognised for notable progress in expanding educational access for women and girls (Layton et al., 2021). The country has managed to ensure a higher level of gender equity in primary school attendance (USAID, 2024). However, such growth often proves futile as most women were found to be unable to complete post-primary education or make the school-to-work transition to good jobs (Layton et al., 2021). Women in Bangladesh continue to be underrepresented in STEM education. Despite recent progress in increasing the number of girls in primary and secondary schools, it is significant to note that, according to 2019 data, only 8% of them ultimately enrol in science departments, with only 1.5% going into engineering and technology (Moyeen, 2023). These numbers are certainly indicative of the gap that exists, and there is a need for immediate action to address it.

Despite remarkable advancements in women's education and microfinance, Bangladesh's female labour force participation rate, including in entrepreneurship, remains low (Chowdhury, 2022). According to the Household Income and Expenditure Survey, 2022, compared to men, Bangladeshi women were less likely to be employed (42.5% vs. 81.3%). Additionally, their unemployment rate was much higher than that of men. Yet, women in Bangladesh have achieved great progress by stepping outside of the house to work and contribute to the economy, especially through entrepreneurship. However, it is important to note that such progress in entrepreneurship is often limited to small and medium-sized enterprises and a few specific industries. It is concerning that Bangladesh remains one of the most challenging countries for women entrepreneurs, as evidenced by its ranking lowest among 58 economies in the Mastercard Index of Women Entrepreneurs for 2020 (Mastercard, 2020). The situation is made worse by systemic obstacles that lead to less involvement of women in entrepreneurship, such as discrimination based on gender, household responsibilities, few networking opportunities, deeply ingrained social norms, and insufficient infrastructure and supportive policies (Wambile et al., 2024).

6.5 WOMEN IN THE TECH INDUSTRY IN BANGLADESH

While Bangladesh is making tremendous strides towards technological advancement, it has been reflected in the growth of women-led e-commerce, especially f-commerce (online business based on Facebook and other social media platforms). F-commerce-based women entrepreneurship was already flourishing, and the COVID-19 pandemic further proved to be a blessing in disguise for many women entrepreneurs in Bangladesh. Due to the pandemic, as more customers were attracted to online shopping, Bangladesh witnessed monumental growth in women-led f-commerce. According to data from Meta, women owned 70% of Facebook businesses opened during the pandemic (Liaquat, 2022). However, it is worth noting here that in such businesses, technology is being utilised just as a platform. Bangladeshi women entrepreneurs' participation in tech startups remains under question. A 2015 report by the International Trade Center highlighted that only 13% of the IT workforce were women and only 1% of ICT companies were led by women (International Trade Center, 2015). Then, after a 3-year gap, a 2018 survey conducted by one of the leading private universities in Bangladesh reported that of the total ICT workforce in Bangladesh, only 16% were women (The Daily Star, 2019). In a country where 50% of the total population is women and where the country is actively looking forward to developing itself as one of the leading digital countries in Asia, the lack of women-led tech ventures is rather eye-catching. Women continue to be underrepresented in numerous IT-related fields and face a higher likelihood of exclusion from novel and emerging positions (Kriton & Robertson, 2018).

6.6 BARRIERS FOR WOMEN IN THE TECH SECTOR IN BANGLADESH

6.6.1 Social Barriers

6.6.1.1 The Idea of Masculinity and Perceived Incapability in the Technology Sector

Women entrepreneurs are disadvantaged by the societal construction of IT as a masculine field because of preconceived cultural and social notions (Saifuddin et al., 2020). This deters women from pursuing technology-related entrepreneurial ventures. Women are frequently viewed as less competent in STEM subjects, further demonstrating how the linking of technical abilities with masculinity fosters an environment where women's competencies are regularly questioned (Cheryan et al., 2017). This prejudice forces women to put in more effort to demonstrate their competence in addition to restricting their options for tech-based entrepreneurial ventures. In Bangladesh, the stereotype persists that men are better suited for the IT sector as men are believed to be more logical, whereas women are perceived as more emotional. Not only does this bias put doubt on women's technical skills, but it also prevents women from having the practical experience that would eliminate these myths. Since men are presumed to be inherently skilled and competent at technical professions, bright women are left on the edges and frequently have to prove themselves repeatedly. Taking a high-tech major during undergraduate studies influences men's careers but not those of women (Saifuddin et al., 2020). Men are still seen as the breadwinner in Bangladesh's traditional, patriarchal society; therefore, many male students feel pressured to major in subjects that will guarantee greater employment prospects and higher-paying jobs. When technical fields are perceived as undesirable for women, it has a substantial impact on women entrepreneurs who want to start enterprises in the tech sector. Women may encounter further obstacles due to this stereotype since they frequently experience mistrust from clients, investors, and even prospective workers, who may question their technical proficiency. Because investors may view female entrepreneurs' businesses as riskier or less feasible than those managed by men, many of them find it difficult to establish credibility and obtain capital.

6.6.1.2 Balancing Work and Family Obligations

Managing work and family obligations can require a lot of time, effort, and commitment from people, which frequently leads to role conflicts. Although balancing these obligations puts pressure on both men and women in the tech sector, stress and burnout are more common among women. Women have additional societal normative restrictions, which hold them responsible for a greater proportion of household chores. In patriarchal cultures like Bangladesh, where women frequently shoulder an excessive share of household duties, this difficulty is even more noticeable. Cultural norms in these contexts support the assumption that a woman's identity or worth as a "good" or "ideal" woman is frequently directly related to her commitment to taking care of her family (Saifuddin et al., 2019). One of the biggest challenges facing women in the tech industry, especially those who want to start their own businesses, is juggling work and family responsibilities. Because of established gender norms, women may be expected to shoulder a greater proportion of household obligations, making the demands of the IT business, which frequently calls for long hours, high availability, and intense devotion, especially stressful. This juggling act might be even more challenging for female would-be business owners because starting a business sometimes involves a significant time commitment that may conflict with obligations to their families.

6.6.1.3 Social Stigma and Lack of Role Models

The above-mentioned barriers as shown in Figure 6.1 highlighted two points. (1) Traditionally, technology-related fields are believed to be intellectually not suitable for women and (2) technology-related fields often demand that individuals

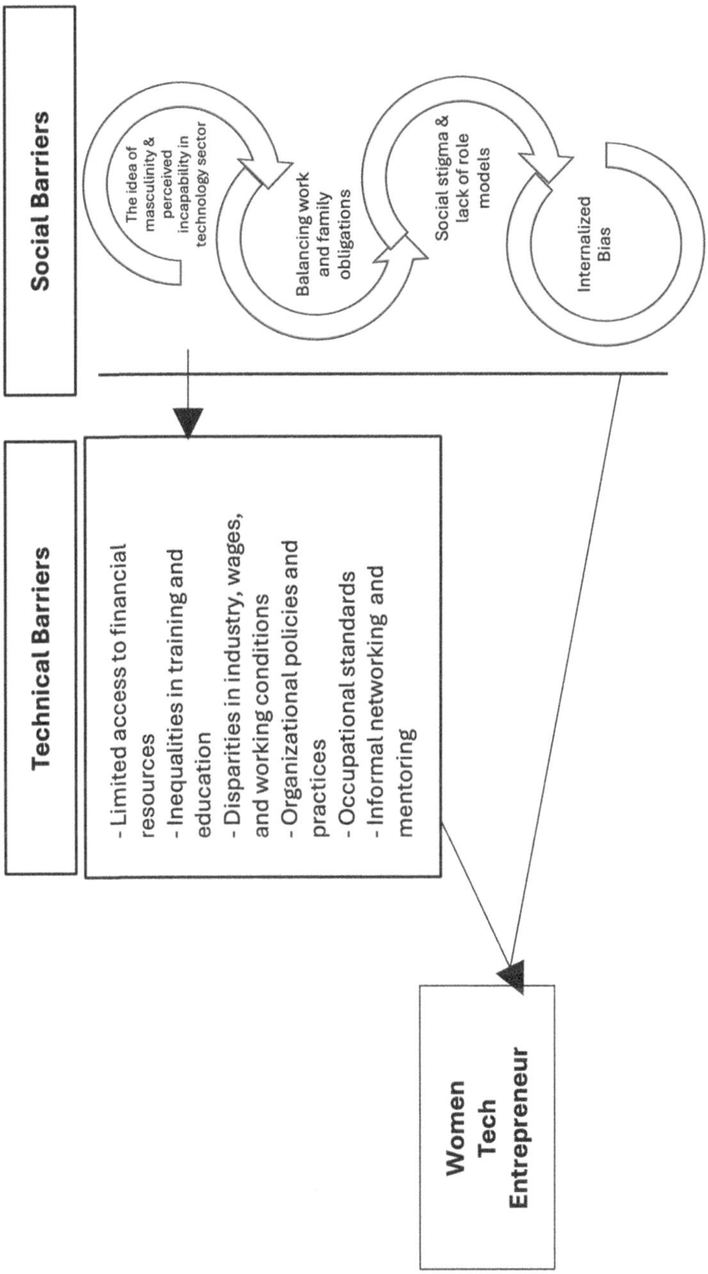

FIGURE 6.1 The Conceptual Model Developed by the Authors.

sacrifice their work-life balance, which makes the field even more challenging for women. That often creates a social stigma around the role of women tech entrepreneurs. Consequently, there has been less participation of women as tech entrepreneurs. Due to the absence of female role models, many female entrepreneurs are discouraged from entering or progressing in the tech sector. This is because they lack visible examples to counteract the gender biases that persist in the industry. Women in the tech industry thus lack inspiration (Danish & Smith, 2012). The lack of well-established female advisers who recognise and support them makes it difficult for women entrepreneurs to establish the networks and credibility necessary for success, which ultimately widens the gender gap in the tech sector (Panda, 2018).

6.6.1.4 Internalized Bias

As the idea of masculinity persists in Bangladesh's society, the above-mentioned social barriers for women often stem from that idea. Consequently, a vicious cycle can be observed that ultimately limits the growth of women tech entrepreneurs. Because the associated social stigma frequently discourages women, both directly and indirectly, from ever considering a career as tech entrepreneurs, many young women find the idea of working in this field intimidating. This leads to the underrepresentation of women in the tech industry, hence sustaining the male-dominated status quo. Negative peer comparisons, ongoing exposure to a male-biased environment, and inefficient teaching strategies tend to cause young women who choose further education in technical subjects to lose interest and self-efficacy (Fisher & Margolis, 2002). And, in this vicious cycle, many women often end up believing that they are not suitable for the tech industry, hence creating an internalised bias.

6.6.2 Technical Barriers

6.6.2.1 Limited Access to Financial Resources

Since investors often perceive women entrepreneurs as less competent or as riskier prospects, they face greater difficulty in securing financing to launch or expand their enterprises. As a result, many women do not have equal access to financial opportunities. This bias is particularly evident in the tech industry, where male-dominated networks frequently influence investment decisions, limiting women entrepreneurs' access to capital and resulting in less favorable funding terms. Such structural inequalities significantly constrain women's ability to establish and scale tech-based ventures.

To expand their firms, women entrepreneurs frequently turn to unofficial financial sources, including friends and family. Their ability to make essential investments in personnel and technology is limited by their reliance on personal savings. Only 7% of venture capital in Bangladesh goes to women-led firms, making it difficult for women entrepreneurs to obtain investment. This is mostly because of prejudices against women entrepreneurs who operate alone and a lack of trust. These challenges are compounded by women entrepreneurs' limited access to collateral for bank loans, under-representation in venture-capital firms, and exclusion from key investment networks, making access to funding a significant barrier (Monamee, 2024).

6.6.2.2 Inequalities in Training and Education

Bangladesh lacks a strong foundation of practical training programmes that are specifically designed to meet the requirements of women, especially in the areas of ICT. Women have limited awareness of the fundamental skills required for the tech industry because many programmes, when they are offered, tend to emphasise academic knowledge over real-world, hands-on experience. When pursuing tech entrepreneurship, women are left with a knowledge gap that affects their confidence and competitiveness since they have not had enough exposure to real-world tools, coding techniques, or problem-solving assignments. Professional codes of ethics and organisational rules should be considered, particularly when it comes to the professional growth and training of a diverse IT workforce (Tapia, 2006). Further supporting the notion that they are underqualified, many women are deterred from seeking technical training or higher education in disciplines like computer science or engineering. This cycle limits varied viewpoints in the IT business, which are critical for innovation and expansion, in addition to limiting women's potential in the field. One of the main driving forces for women's empowerment has been IT education. It is challenging to provide even the most basic ICT training in rural areas, where more than 70% of Bangladesh's population still resides (Bangladesh Bureau of Statistics, 2022). Women still face major obstacles in entering the technology industry due to disparities in education and training, particularly those who want to start their own business. Because of prejudices that push women to seek traditionally "female" careers rather than pursue STEM subjects, women in many areas have fewer options for technical education. Women find it more difficult to gain the fundamental knowledge and expertise required to thrive in digital entrepreneurship due to limited access to mentorship, development courses, and tech-focused training programmes. Women must overcome several obstacles simply to acquire entry-level computing, coding, and technical resource skills, whereas men frequently profit from early exposure to these subjects (Tondon, 2006). Due to social and household responsibilities, women frequently have restricted time and fewer possibilities for practical training, which is essential for developing ICT competency. Furthermore, balancing several responsibilities like taking care of the home, providing care, and pursuing career goals leaves women with little time and energy to invest in skill development. Systemic barriers prevent women from reaching their full potential in ICT, even when they possess the motivation and aspirations, as a result of this "time poverty" which further limits their capacity to progress in technological careers.

6.6.2.3 Disparities in Industry, Wages, and Working Conditions

Although technology is changing the job market in Bangladesh and creating new opportunities for tech-savvy, higher-paying positions, women are still overrepresented in traditionally labour-intensive, lower-skilled fields (Tapia, 2006). These positions rarely provide opportunities for skill development or advancement and are frequently characterised by poor pay and little room for advancement. As a result, women have fewer opportunities for career progression while men stand to gain from the changing digital landscape, leading to a growing economic divide. The loop is further strengthened by cultural norms that direct women towards these low-growth sectors, diminishing their prominence in the tech industry and sustaining the idea that high-tech employment is "unsuitable" for women. This disparity not only limits women's earning potential but also denies the nation access to a fully qualified, diverse workforce that may drive innovation in new technological fields. Women in technology face significant obstacles due to disparities in the industry, pay, and working conditions, which affect both current employees and prospective business owners. Even when they have the same credentials and do the same work, women in tech positions sometimes see large pay disparities in comparison to men. Women are less able to start or maintain tech-based firms as a result of this salary gap, which also limits the amount of money they can reinvest in entrepreneurial endeavours. In the tech industry, women make about 70% of what males do, according to the World Economic Forum (2021). In addition, women frequently experience harassment and discrimination at work, which leads to high turnover rates and restricts their prospects for promotion.

6.6.2.4 Organisational Policies and Practices

Senior management makes most of the decisions, and males typically occupy those roles. They attempt to give preference to men, whether on purpose or unintentionally. This gender imbalance at the top creates a subtle but powerful bias, where decisions naturally align with the experiences and preferences of male leaders. As a result, policies around career advancement, work-life balance, and even skill-building programmes tend to be structured in ways that may not account for the unique challenges faced by women. This dynamic creates a corporate culture that, intentionally or unintentionally, reinforces male dominance and limits the inclusivity and flexibility necessary to attract more women to and retain them in the workforce, especially in technology-related roles (Saifuddin et al., 2019). Women's professional advancement and entrepreneurial success in the technology sector are impeded by gender inequities that are frequently reinforced by organisational rules. Fewer women hold technical and leadership positions in many digital organisations because of hiring, promotion, and retention policies that unintentionally favour male workers. Policies that downplay the value of diversity or that do not provide for balance between work and life, like parental leave or flexible work options, can deter women from pursuing long-term tech jobs or from accumulating the knowledge required to launch their own companies. Women are disproportionately impacted by the lack of family-friendly policies such as flexible work schedules and parental leave in many tech organisations. This lack of assistance may result in increased employment turnover among female entrepreneurs.

6.6.2.5 Occupational Standards

Due to the high-tech industry's extremely demanding requirements, individuals must frequently do complex, flexible activities outside of regular business hours. In addition to these long hours, the requirement for ongoing skill improvements creates a demanding work environment. Because women frequently feel pressured to put their home responsibilities first, Bangladeshi women in formal employment find it challenging to dedicate themselves fully to or progress in the high-tech sector (Saifuddin et al., 2019). Women face additional challenges due to occupational conventions, which frequently subject them to family and social constraints resulting from patriarchal mindsets and prevailing social norms (Srinivasan et al., 2013). In the tech sector, workplace rules often unintentionally favour men, which puts more obstacles in the path of women, especially those who want to start their own businesses. Women who may have other obligations outside of work may find it challenging to meet these expectations, which usually emphasise long hours, continuous availability, and an aggressive work ethic. These expectations, which have their roots in a culture that has historically been dominated by men, might deter women from accepting or staying in tech positions, which limits their chances of gaining the expertise and knowledge of the field needed for tech entrepreneurship.

6.6.2.6 Informal Networking

The absence of possibilities for informal networking is another major obstacle for women's tech entrepreneurship (Saifuddin et al., 2019). These kinds of networks are essential for exchanging information on recent advancements in the industry and acquiring financing or investments. Women are frequently unable to attend these activities due to family constraints. They thereby lose out on making important connections that are necessary for acquiring financial investments and social capital for their entrepreneurial ventures. Informal networking can help individuals learn about opportunities and build relationships (Forret & Dougherty, 2004). Most of these informal networks are male-dominated, and women are frequently left out, particularly those engaged in tech entrepreneurship. It can be difficult for women to attend social parties, tech meetups, and informal after-hours activities because they feel uncomfortable in these male-dominated settings or because they have personal and familial obligations that prevent them from going. Due to their limited access to informal networking, women entrepreneurs may find it difficult to establish important connections with partners, investors, and other industry participants. Women find it difficult to become visible without these relationships which aid male business owners in overcoming obstacles and growing their enterprises. Moreover, if women-led firms do not have established connections, investors and possible partners might be less inclined to help them because these unofficial networks frequently affect decision-making and trust. Women in technology have trouble connecting with mentors, investors, and colleagues in the business since they cannot easily access informal networks.

6.6.2.7 Mentoring

As a key tactic for entrepreneurial development, mentoring is essential, especially in the tech industry. However, establishing these mentoring relationships can be challenging for women tech entrepreneurs. Concerns about possible career consequences, such as being linked to an unsuccessful mentee or facing accusations of impropriety, could discourage senior male entrepreneurs or tech experts from mentoring women entrepreneurs. There are times when senior male colleagues are unwilling to mentor women entrepreneurs out of concern that being linked to a failing female protégé could harm their career. It is especially challenging for female entrepreneurs to locate a female mentor due to the lack of women in the tech industry occupying senior roles. Because of this, women frequently lose out on the assistance and direction that men easily obtain, which restricts their ability to grow and progress in their entrepreneurial ventures (Saifuddin et al., 2019). Using interviews with 50 senior female managers, Linehan and Scullion (2008) investigate the effects of networking and mentorship on the professional growth of female global managers. Their experiences highlight issues with unofficial organisational procedures that impede professional growth. Female managers frequently lose out on international possibilities because they lack access to networks, mentors, role models, and sponsorship that their male counterparts have. According to Cross and Linehan (2006), women occasionally do not receive the same level of mentoring as their male counterparts. For women in the tech industry, mentoring is essential, especially for those who want to start their own businesses. Women who have access to good mentorship connections can gain important advice, encouragement, and insights into negotiating the intricacies of the tech sector. However, because women are underrepresented in the tech industry, many female would-be entrepreneurs find it difficult to locate mentors who can relate to their women-specific struggles and provide helpful guidance. Women who have access to mentorship are more likely to seek jobs in entrepreneurship and technology. However, a lot of ambitious female entrepreneurs have trouble finding mentors who can offer support and guidance because there aren't many female role models in top positions.

6.7 CONCLUSION

To sum up, the relationship between gender, technology, and entrepreneurship in Bangladesh highlights important obstacles that prevent women from entering the tech sector. Women are discouraged from pursuing entrepreneurship in the tech industry because of cultural beliefs that are entrenched in traditional masculinity and frame technology as a male realm. This is made worse by structural disparities in education, restricted access to mentorship and networks, and a lack of family-friendly workplace regulations. The present educational system, which mainly emphasises cognitive growth, needs to be changed to give people the practical skills required in the modern workforce, particularly in the aftermath of the COVID-19 pandemic (Shahriar et al., 2021). Additionally, infrastructure issues impede the effectiveness of digital education (Aziz, 2020). It is essential to advance ICT education, dispel prejudices, and put supportive organisational policies into place to create an inclusive digital workplace. By empowering women entrepreneurs and utilising their varied viewpoints, the industry can stimulate economic expansion and adjust to the changing needs of a quickly shifting environment.

REFERENCES

Afroja, A., & Islam, M.S. (2023). Moving towards digital trade transformation in Bangladesh: prospects and challenges for implementing digital trade. *Journal of Business Strategy Finance & Management*, *5*(2), 115–142.

Akter, S., Amin, M. A., Hossain, M. N. U., Yasmin, A., Sarker, M. N. I., Ali, I., & Azman, A. (2024). The impact of the Fourth Industrial Revolution on female workers in the ready-made garments and textile industry, Bangladesh. *The International Journal of Community and Social Development*, *6*(3), 317–338.

Alam, M. S., Ali, M. J., Bhuiyan, A. B., Solaiman, M., & Rahman, M. A. (2020). The impact of Covid-19 pandemic on the economic growth in Bangladesh: a conceptual review. *American Economic & Social Review*, *6*(2), 1–12.

Alam, M. J., & Ogawa, K. (2024). Future-readiness and employable ICT skills of university graduates in Bangladesh: an analysis during the post-covid era. *Cogent Social Sciences*, *10*(1), 1–7.

Ashaari, M. A., Singh, K. S. D., Abbasi, G. A., Amran, A., & Liebana-Cabanillas, F. J. (2021). Big data analytics capability for improved performance of higher education institutions in the era of IR 4.0: A multi-analytical SEM & ANN perspective. *Technological Forecasting and Social Change*, *173*, 121119.

Azran, A.(2024, February 15) The rapid growth of Bangladesh's economy, and what comes after. *The Daily Star.* Available at: www.thedailystar.net/anniversary-supplement-2024/bangladesh-the-world-stage/news/the-rapid-growth-bangladeshs-economy-and-what-comes-after-3543361 [Accessed 24 October 2024]

Aziz, A. (2020). Digital inclusion challenges in Bangladesh: the case of the National ICT Policy. *Contemporary South Asia*, *28*(3), 304–319.

Bangladesh Bureau of Statistics (2018). Labor force survey- Bangladesh 2016-2017. Dhaka.

Bangladesh Bureau of Statistics. (2022). Population and housing census 2022: preliminary report (English). Available at: https://sid.portal.gov.bd/sites/default/files/files/sid.portal.gov.bd/publications/01ad1ffe_cfef_4811_af97_594b6c64d7c3/PHC_Preliminary_Report_(English)_August_2022.pdf

Bangladesh High Tech Park Industry (2024). *Data on tech industry.* Available at: https://new.bhtpa.gov.bd/data-on-the-tech-industry/n [Accessed 23 October, 2024]

Bangladesh High Tech Park Authority (2024). About us. Available at: https://ossbhtpa.gov.bd/about-us [Accessed 22 October 2024]

Bangladesh Investment Development Authority (n.d.). Information technology – BIDA overview of Bangladesh IT/ITES sector. Available at: https://boi.gov.bd/information-technology

Bangladesh Investment Development Authority (2019). *Information technology industry.*

Bangladesh Investment Development Authority (2021). *Telecommunication & networking sector.*

Bangladesh Telecommunication Regulatory Commission (2024a). *Mobile subscriber.* Available at: https://btrc.gov.bd/site/page/0ae188ae-146e-465c-8ed8-d76b7947b5dd/- [Accessed 24 October 2024]

Bangladesh Telecommunication Regulatory Commission (2024b). *Internet subscriber.* Available at: https://btrc.gov.bd/site/page/347df7fe-409f-451e-a415-65b109a207f5/- [Accessed 24 October 2024]

Barry, J. J. (2018). *Information communication technology and poverty alleviation: Promoting good governance in the developing world.* Routledge.

Basharat, L., & Alam, M.J. (2024). Workplace environment for gender equality and sustainable career planning: the case of Bangladesh. *People, Spaces and Places in Gendered Environments*, *34*, 77–97.

Basole, R. C., & Patel, S. S. (2018). Transformation through unbundling: visualizing the global fintech ecosystem. *Service Science*, *10*(4), 379–396.

The Business Standard (2023, August 24). *Women in start-up: a journey towards Smart Bangladesh.* Available at: www.tbsnews.net/supplement/women-start-journey-towards-smart-bangladesh-688162 [Accessed 24 October 2024]

Cheryan, S., Ziegler, S. A., Montoya, A. K., & Jiang, L. (2017). Why are some STEM fields more gender balanced than others? *Psychological Bulletin*, *143*(1), 1.

Chowdhury, A. (2022). *Progressing female labour force participation in Bangladesh.* International Growth Center. Available at: www.theigc.org/blogs/gender-equality/progressing-female-labour-force-participation-bangladesh [Accessed 10 October 2024]

CIA World Factbook (2024). *Country comparisons: telephones-mobile cellular.* Available at: www.cia.gov/the-world-factbook/field/telephones-mobile-cellular/country-comparison/ [Accessed 15 October 2024]

Cross, C., & Linehan, M. (2006). Barriers to advancing female careers in the high-tech sector: empirical evidence from Ireland. *Women in Management Review, 21*(1), 28–39.

Danish, A. Y., & Smith, H. L. (2012). Female entrepreneurship in Saudi Arabia: opportunities and challenges. *International Journal of Gender and Entrepreneurship, 4*(3), 216–235.

The Daily Star (2019, March 6). *Of the ICT workforce, only 16% are women: study.* Available at: www.thedailystar.net/country/ict-sector-in-bangladesh-only-16-per-cent-are-women-study-1711279 [Accessed 24 October 2024]

The Daily Star (2024, February 7). *Bangladesh top in global software development growth: GitHub report.* Available at: www.thedailystar.net/tech-startup/news/bangladesh-top-global-software-development-growth-github-report-3538621 [Accessed 24 October 2024]

Fisher, A., & Margolis, J. (2002). Unlocking the clubhouse: the Carnegie Mellon experience. *ACM SIGCSE Bulletin, 34*(2), 79–83.

Forret, M. L., & Dougherty, T. W. (2004). Networking behaviors and career outcomes: differences for men and women? *Journal of Organizational Behavior: The International Journal of Industrial, Occupational and Organizational Psychology and Behavior, 25*(3), 419–437.

Gërguri-Rashiti, S., & Rotabi, S. (2021). Women entrepreneurship in developing countries: with focus on GCC and Kuwait. In *The Emerald handbook of women and entrepreneurship in developing economies* (pp. 91–109). Emerald Publishing Limited.

Hasan, I., Khan, S. N., & Kamal, N. (2024). E-entrepreneurship prospects and women empowerment: a case presenting women entrepreneurs. *Tuijin Jishu/Journal of Propulsion Technology, 45*(3), 736–743.

Hossain, D. M., Alam, M. S., Mazumder, M. M. M., & Amin, A. (2021). Gender-related discourses in corporate annual reports: an exploratory study on the Bangladeshi companies. *Journal of Accounting & Organizational Change, 17*(3), 394–415.

Hossain, N. (2021). The SDGs and the empowerment of Bangladeshi women. In S. Chaturvedi, H. Janus, S. Klingebiel, X. Li, A. de Mello e Souza, E. Sidiropoulos, & D. Wehrmann (Eds.), *The Palgrave handbook of development cooperation for achieving the 2030 agenda: contested collaboration* (pp. 453–474). Palgrave Macmillan. https://doi.org/10.1007/978-3-030-57938-8_21

International Trade Center (2015). *Bangladesh: overcoming gender challenges in the IT sector.* Available at: www.intracen.org/fr/node/1287 [Accessed 24 October 2024]

Islam, M. N., & Inan, T. T. (2021). Exploring the fundamental factors of digital inequality in Bangladesh. *Sage Open, 11*(2) 1-12. doi.org/10.1177/21582440211021407.

Islam, M. R. (2019a). Economic growth rates and exports of Bangladesh: the Bengal tiger? *South Asia Research, 39*(3), 285–303.

Islam, M. Z. (2019b, December 15). Telecom sector's revenue to cross $5b by 2023. *The Daily Star.* Available at: www.thedailystar.net/business/news/telecom-sectors-revenue-cross-5b-2023-1840516 [Accessed 24 October 2024]

Islam, Q. T., Zabir, F., Asif Hossain, M., & Iqbal, R. (2023). Challenges of adapting to the Fourth Industrial Revolution in emerging economies: a Bangladesh CASE. In *Industry 4.0 and the Digital Transformation of International Business* (pp. 21–32). Springer Nature Singapore.

Kabir, S. (2023, March 7). Bangladeshi women in tech and digital education: challenges and strides. *The Daily Star*. Available at: www.thedailystar.net/supplements/international-womens-day-2023/news/bangladeshi-women-tech-and-digital-education-challenges-and-strides-3266051 [Accessed 24 October 2024]

Kamarthi, S., & Li, W. (2020). Technology enablers for manufacturing resilience in the COVID-19 and post–COVID–19 era. *Smart and Sustainable Manufacturing Systems*, *4*(3), 20200064.

Khatun, F., Saadat, S.Y., & Islam, M.J. (2021). *Digital platform economy in Bangladesh: opportunities and challenges*. Centre for Policy Dialogue and Friedrich-Ebert-Stiftung Publication.

Kirton, G., & Robertson, M. (2018). Sustaining and advancing IT careers: women's experiences in a UK-based IT company. *The Journal of Strategic Information Systems*, *27*(2), 157–169.

Lata, L. N., Walters, P., & Roitman, S. (2021). The politics of gendered space: social norms and purdah affecting female informal work in Dhaka, Bangladesh. *Gender, Work & Organization*, *28*(1), 318–336.

Layton, L.A., Lusk-Stover, O., Inoue, K., & Sosale, S. (2021). How to provide opportunities for all? From girls' education to women's labor force participation in Bangladesh. The World Bank. Available at: https://blogs.worldbank.org/en/education/how-provide-opportunities-all-girls-education-womens-labor-force-participation-bangladesh [Accessed 24 October 2024]

Liaquat, Z.B. (2022, March 8). Women led business boom during pandemic. *Dhaka Tribune*. Available at: www.dhakatribune.com/business/commerce/265172/women-led-business-boom-during-pandemic [Accessed 24 October 2024]

Linehan, M., & Scullion, H. (2008). The development of female global managers: the role of mentoring and networking. *Journal of Business Ethics*, *83*, 29–40.

Mahmud, K., Joarder, M. M. A., & Sakib, K. (2023). Customer fintech readiness (CFR): assessing customer readiness for fintech in Bangladesh. *Journal of Open Innovation: Technology, Market, and Complexity*, *9*(2), 100032.

Massally, N. K., Khurshed, N., & Tellez-Merchan, N. (2022). Measuring progress to scale: responsible digital payment in Bangladesh.

Mastercard (2020). The Mastercard index of women entrepreneurs 2020 report.

Monamee, M. I. (2024, March 8). Breaking barriers: unveiling the challenges faced by women entrepreneurs. *The Daily Star*. Available at: www.thedailystar.net/tech-startup/news/breaking-barriers-unveiling-the-challenges-faced-women-entrepreneurs-3562001

Moyeen, S. (2023, June 20). *Bangladesh's women rise to the challenge with big ideas for digital innovation*. End Poverty in South Asia. The World Bank. Available at: https://blogs.worldbank.org/en/endpovertyinsouthasia/bangladeshs-women-rise-challenge-big-ideas-digital-innovation [Accessed 24 October 2024]

Mujeri, M. K., & Mujeri, N. (2020). *Bangladesh at fifty: moving beyond development traps*. Palgrave Macmillan.

Panda, S. (2018). Constraints faced by women entrepreneurs in developing countries: review and ranking. *Gender in Management: An International Journal*, *33*(4), 315–331.

Rahman, A., & Abedin, M. J. (2021). The Fourth Industrial Revolution and private commercial banks: the good, bad and ugly. *International Journal of Organizational Analysis*, *29*(5), 1287–1301.

Saifuddin, S. M., & Baker, S. D. (2020). A conceptual model of leadership aspiration using theory of planned behavior. *Applied Management Journal*, *21*.103–130

Saifuddin, S., Dyke, L., & Hossain, M. S. (2019). Walls all around: barriers women professionals face in high-tech careers in Bangladesh. *Equality, Diversity and Inclusion: An International Journal*, *38*(7), 705–726.

Shahriar, M. S., Islam, K. M., Zayed, N. M., Hasan, K., & Raisa, T. S. (2021). The impact of COVID-19 on Bangladesh's economy: a focus on graduate employability. *The Journal of Asian Finance, Economics and Business*, *8*(3), 1395–1403.

Srinivasan, V., Murty, L. S., & Nakra, M. (2013). Career persistence of women software professionals in India. *Gender in Management: An International Journal*, *28*(4), 210–227.

Sultana, S.A. (2021, May 26). *Women entrepreneurs in Bangladesh: barely staying afloat*. The Asia Foundation. Available at: asiafoundation.org/women-entrepreneurs-in-bangladesh-barely-staying-afloat/

Tabassum, N., & Nayak, B. S. (2021). Gender stereotypes and their impact on women's career progressions from a managerial perspective. *IIM Kozhikode Society & Management Review*, *10*(2), 192–208.

Tandon, N. (2006). *Information and communication technologies in Bangladesh: trends, opportunities and options for women workers*. USAID.

Tapia, A. H. (2006). Hostile work environment.com: increasing participation of underrepresented groups, lessons learned from the dot-com era. *ACM SIGMIS Database: The Database for Advances in Information Systems*, *37*(4), 79–98.

Upturn Bangladesh (n.d.). Report on the information technology industry in Bangladesh. https://www.upturn.com.bd/industry/it

USAID (2024). Gender equality and women's empowerment.

Wambile, A., Bonfert, A.T., Kalle, A., Moylan, H., & Muller, M. (2024). What do gender data reveal about the economic struggles of women in Bangladesh? The World Bank. Available at: blogs.worldbank.org/en/opendata/what-do-gender-data-reveal-about-economic-struggles-women-bangladesh [Accessed 24 October 2024]

World Bank (2024). *The World Bank in Bangladesh-overview.* Available at: www.worldbank.org/en/country/bangladesh/overview [Accessed 24 October 2024]

World Economic Forum (2021). *Global gender gap report 2021* (Geneva, Switzerland). World Economic Forum. Available at: https://www3.weforum.org/docs/WEF_GGGR_2021.pdf

7 The Impact of Women Empowerment on the Economic Development of Bangladesh

An Empirical Analysis

Anta Atalantia, Mostafizur Rahman, and Rifah Nanjiba

7.1 INTRODUCTION

"The empowerment of women is one of the central issues in the process of development for many countries in the world today" (Sen, 1999b, p.202).

Nobel laureate Amartya Sen's quote highlights the significance of women empowerment for a country's development. He argues that one of the primary forces and a key component of the development process is changing women's agency. This chapter explores the economic route to economic development through women's empowerment in Bangladesh, where women's lifelong dependence on males and their lack of financial and educational prospects have long been considered the foundations of their inferior status (Kabeer, 2017). Bangladesh has achieved relatively fast progress towards gender equality since its liberation from Pakistan in 1971, and now it is catching up and even outpacing regional indicators of the Human Development Index (HDI) on life expectancy, education, and health, with an annual gross domestic product (GDP) growth rate of 7.9% (World Bank, n.d.). The total population of Bangladesh was 168 million in 2022, where 49.5% of the total population was female. However, there is an enormous gender gap in the labour force and educational attainment in Bangladesh. Even though they make up half of the population, women are still an underutilised resource, constituting only 30% of Bangladesh's workforce (World Bank, 2021). According to Sen, a woman's well-being is closely related to her capacity to earn independently, engage in work outside the house, own substantial inheritance ownership rights, be literate, and make decisions both inside and outside the household (Sen, 1999b). Sen's core arguments assert the dialectical relationship between development and freedom. In essence, "Inequality between women and men prematurely ends the lives of millions of women, and, in different ways, severely restricts the substantive freedoms that women enjoy" (Sen, 1999b, p.15).

DOI: 10.1201/9781003544838-7

He claims that these freedoms are access to healthcare, education, political dissent, free markets, and equality which all promote the development of one another. Also, he opines that there is a positive relationship between gender inequality and development. This relationship primarily functions through women's agency, including female labour force participation, female literacy, and fertility rate. The influences like women's literacy and women's employment can create a massive change in fertility rates (Sen, 1999a). So, following the enlightenment of previous literature, this chapter captures the dimensions of women's empowerment and autonomy through three main variables: female literacy rate, fertility rate, and female labour force participation, and explores the relationship between women empowerment and economic development. Economic development and women's empowerment are closely related. In one way, economic development can lessen gender disparity. In another way, women's empowerment can be one of the main driving factors for economic development (Duflo, 2012). Poverty and lack of opportunities create gender inequality. On the one side, economic development reduces poverty and gender inequality. As a consequence, women's condition improves more than men. On the other hand, women's empowerment causes economic development. However, economic development alone cannot bring total equality between men and women (Duflo, 2012). Thus, to add to this growing stream of studies, this study will empirically examine the following research question: Is there any cointegrating and causal relationship between women empowerment (fertility rate, female labour force, and female literacy rate as the indicators) and economic development (as a measure of per capita GDP and HDI) in Bangladesh? This study contributes to the research by using an inclusive measure of women's empowerment by measuring three main affecting variables (female labour force, female literacy, and fertility rate); also to measure its impact on economic development, the study estimates both GDP and HDI of the country. Conferring to Sen (1999a), development is seen as freedom. Sen elucidated that the measurement of development only by the growth of GNP or GDP or rise in personal income or industrialisation will narrow the views of development aspects. According to him, development also depends on other factors like access to education, life expectancy, and freedom of speech. Thus, in this study, the HDI also with GDP is taken to measure economic development comprehensively. Findings based on the autoregressive distributed lag model (ARDL) and the Granger causality test show significant cointegration and causal relationship between women empowerment and economic development.

7.1.1 Gender Gap

It is common to use the terms "gender gap", "gender inequality", and "empowerment" simultaneously. The gender gap is often measured as wage inequality, gender-based legal rights, and educational gap, which often compares women's position to men. In developing countries, gender inequality or women's inferior status has been recognised as an absolute obstacle to development. Empowering women is a prerequisite for a country's growth and development (Sen, 2001). On the Global Gender Gap Index 2022, Bangladesh scored 0.714, placing it in 71st position out of 146 countries (World Economic Forum, 2022).

7.1.2 Why Reducing Gender Gap and Empowering Women Is Important

There is a visible hesitancy to consider gender an independent concept in development literacy (Sen, 1987). Sen describes it as misleading to study only income and class categories for some problems. The gender-based analysis is a vital consideration in societal and economic analysis and is complementary to the variable of employment, family status, and income. Women's steady inferior status in society refers to the necessity of considering gender as a separate development aspect. Women have biological advantages in survival and longevity when they get the basic need of living, such as nutrition and health care. Sen emphasised that while there are 106 women for every 100 men in Europe and North America, the least-developed countries only have 97 women for every 100 men. China's missing number of women is 38 million (Sen, 1987).

7.1.3 Definition and Measurement of Women's Empowerment

Sundström et al. (2017) described women's empowerment as the ability to make a better choice, agency, household, and social decision-making, which is the universal standard term than the relative term of women-to-men in society. Kabeer (2005) defines empowerment through three correlated dimensions. Firstly, agency is the process through which strategic life choices are made. Secondly, empowerment refers to not only the decision-making process by individuals but also their sense of agency and self-esteem. Thirdly, resources refer to which agency is distributed through institutions and relations in a society. The outcome of resources and agency is the achievements exercised as the outcome of individual efforts to attain empowerment. Kabeer (2005) argues that educated women in rural Bangladesh participate more in household decision-making than uneducated ones. Findings support that literacy enhances self-esteem in rural Bangladesh (Hashemi et al., 1996). In comparison to many countries, women have more involvement in politics and government programs with wider self-esteem and confidence in public affairs for longer-term memberships with micro-credit organisations in Bangladesh. Women workers in the Ready Made Garments (RMG) sector of Bangladesh are more satisfied having a suitable job than in other poorly paid informal job sectors (Kabeer, 2005). Haque et al. (2011) examine women's empowerment through three dimensions: financial decision-making power, household bargaining power, and liberty in physical movement in the context of Bangladesh. They refine the differences between the concepts of women empowerment, women's status in society, autonomy, and gender disparity in Bangladesh. Conferring to the study, women's participation in decision-making regarding financial matters reduces their economic degradation in society. Also, the three primary indicators of the economic decision-making process, that is, taking important financial decisions, making daily household purchases, and sharing opinions on spending money, can measure women's empowerment. Participating in domestic affairs like their own health care, their child's health care, food consumption, and family planning improves women's self-esteem, autonomy, and social status. Similar to the previous study, Hashemi et al. (1996) indicate eight dimensions to measure women empowerment. The significant indicators according to this study are freedom of physical movement, household and economic bargaining power, participation in the political sector, and citizen awareness. Kabeer (2017) explores two pathways to conceptualise women's empowerment. One way is to evaluate what can change women's lives from their priorities and understandings. Another way is to assess external determinants to empower women in a patriarchal societal context. The first approach will show how they infer and evaluate their perspective of social change and empowerment within themselves. The second approach will determine the unequal ability to make choices between women and men in society. An individual's power (Dahl, 1957), autonomy, or empowerment does not reflect the ability to make choices but rather the availability of choices that are perceived as possible to acquire. Here, Kabeer analysed how she perceives the concept of "power" as it draws the root concept of empowerment. Based on the standard sociological definition, the ability to make plenty of choices individually and execute the choices on others is regarded as power (Dahl, 1957), but according to Kabeer, an individual is not empowered if they were not previously disempowered. Empowerment refers to the process of change through which one gains the ability to make choices from which they have been deprived. To define choice in the context of women's empowerment, it should be considered in what conditions the choice is made, and to make a choice means there should be the ability to take alternative choices. If a woman cannot behave or think outside of the prescribed conventional social norms, then this behaviour cannot be considered freedom of choice. Kabeer elucidates that we must differentiate the concept of taking strategic choices, which signifies taking control of our own life choices to lead our lives in a way we want them to be, to make the change happen. For example, Kabeer said there is a distinction between being able to choose to buy soap and deciding whom to marry. So, empowerment refers to the ability to make tactical and important choices by individuals deprived of this ability and challenge the structural norms of inequality in society.

7.2 LITERATURE REVIEW

Although educational accomplishment is essential for women's empowerment, families devote more importance to secondary education for boys than girls in Bangladesh. Girls are more likely to involve in domestic tasks and income sources, and early marriages result in higher school absenteeism and dropping out of school (Dutta, 2025). Husbands' education also needs to understand wives' freedom of choices about their own lives and the family decision-making process. A survey result from the study by Haque et al. (2011) show that women with higher education have higher control over economic and household bargaining processes. But uneducated women have the most increased autonomy in physical movement. This suggests that educational attainment in a patriarchal society like Bangladesh does not seem to improve all spares of empowerment and autonomy of women even though education is regarded as a human investment which can result in higher economic productivity (Hossain and Tisdell, 2005). The human capital theory implies that an equal level of education can determine equal pay and employment (Becker, 1993). The social theory also suggests that more educated women tend to have a lower fertility rate. Also, female literacy negatively impacts the infant mortality rate (Schultz, 1989). Macro-level data analysis of Hossain and Tisdell (2005) shows that labour force participation and educational attainment have increased in Bangladesh, and this would promote women's autonomy through knowledge and information. The study also shows a positive correlation between female labour force participation and female literacy. Results indicate that the low-skill manufacturing sector has been a prime source of female employment. Also, Kabeer and Huq (2014) shows that educated rural women in Bangladesh have a wide range of participation in household decision-making. A study by Sen suggests that educated women in West Bengal, by increasing self-esteem among them, are better able to deal with domestic violence than uneducated women (Sen, 1999a). Similar findings from Bangladesh support this result (Riley et al., 1996). Aslam (2014) explains that for equalising education for women, there have to be lower barriers to their opportunities. The paper studies the impact of education on social, gender, and economic empowerment. Firstly, the study uses household and labour force data from Bangladesh, Pakistan, and India to assess the correlation between women's education and employment. Besides measuring educational attainment and employment, the study measures other indicators of women empowerment: control over financial resources, awareness about public affairs, and their perceptions of empowerment. The study findings show that educational background works as a vehicle to empower women, improving the chances of women entering the labour market. On the other hand, culture and norms significantly constrain the path of higher education for women.

Kabeer et al. (2017) elucidate the differentiation between paid and unpaid labour by women and how this relates to women's empowerment. According to Kabeer, we will set the barrier for measuring empowerment very low if we consider any "choice" as empowerment. She emphasises the differences in outcomes from different labour forces for women. UN women's development study examined a survey of different kinds of work activities undertaken by women in eight districts of Bangladesh (Kabeer et al., 2013). Women's work was categorised into six economic activities: formal, semi-formal, or informal paid work, self-enterprise within and outside households, unpaid work, and financial inactivity. The research also includes other income sources, such as landowner, non-government organisation (NGO) membership, and microfinance programs, which also feature the educational background. This study reflects the multi-dimensional conceptualise approach to measuring women's empowerment which shows that formal or semi-formal works have the most favourable outcome than the rest of the work in measuring empowerment. Also, paid work outside the house offers a more consistently positive relationship with empowerment than paid work inside the home. Interestingly, informal paid working women are less empowered than economically inactive women. And education proved to be the most critical factor than land ownership. In conclusion, she expounds that women's participation in work, whether inside the household or outside of it, whether by micro credit loans or other market forces, is not enough to uphold active citizenship and political agency. Anderson and Eswaran (2009) examined rural households and compared women who earned by themselves with household poultry to those who were economically inactive, working as unpaid workers in the household, or on their husbands' farms. The results show that women engaged in self-enterprise had more power in decision-making than those who were economically inactive. Educational level did not affect this, but financial assets improved their purchasing power in the family. Salway et al. (2005) studied married women in a slum near Dhaka in Bangladesh. They compared urban waged working women, who mainly worked in the garments factory, construction work, and domestic help service, to those women who are economically inactive. The findings suggest that working women have more say in financial management, have more liberty of physical movement, and are less likely to have domestic violence than unemployed women. Heath (2014) examined garments working women and found that domestic violence was higher with paid women workers than with unpaid women workers. Still, these results confined workers with lower education and early marriage. So, irrespective of engagement in garments work, women with higher education faced lower domestic violence. Kabeer compared the findings of two studies, noting that women who reported experiencing domestic violence in Salway et al. (2005) were also among those who later took up waged employment in Heath (2014). This overlap suggests a positive association between women's participation in paid work and their exposure to domestic violence.

Building on the reputation and popularity of Mohammad Yunus, who pioneered microcredit initiatives targeting women in Bangladesh, numerous studies have examined the relationship between microfinance and women's empowerment. These studies mainly evaluate access to microcredit, relatable training, and the course of membership in the women's group. Riley et al. (1996) studied two programs in Bangladesh—the Bangladesh Rural Advancement Committee (BRAC) and Grameen Bank, that offer microfinance to the rural poor of Bangladesh. These programs offer financial services to poor people on a large scale with a low and diminishing scale of financial subsidies. Critics of these programs argue that these minimalist programs do not empower women but worsen their situation. They argue that the patriarchal social structure should be changed rather than providing micro-credit loans to women for their empowerment. Riley et al. (1996) analysed survey data of the two programs on eight dimensions to empower women. The study results show that the longer women remain as a member of each micro-credit program, the greater the potential to be empowered based on the authors' eight indicators of women empowerment. The credit program has significant effects on the indicators of empowerment as it induces economic empowerment in the family, women's contribution to the family, and their political and legal awareness, which also support the findings of Copestake et al. (2005). The ethnographic research finding of the study suggests that through the grouping process, women get access to the outside world more, and they learn to talk to others; the programs increase women's mobility within and outside of the villages. This way, women can fight gender barriers and reduce household violence. Kabeer (2001) finds that the Small Enterprise Development Program (SEDP), which offers moderate-sized loans to women and men, reduces violence against women significantly. Also, extended membership with BRAC and NGOs increases women's political awareness. Debnath et al. (2019) studied micro-credit accessibility by rural women of Bangladesh and how microcredit induces empowerment among them. The empirical findings state that affiliation with micro-credit programs and borrowings increase their ability to control their savings but did not significantly improve their ability to control their income. Additionally, access to micro-credit increases five dimensions of women empowerment: financial stability, physical movement, purchasing power, household decision-making, and political awareness.

Nobel laureate Amartya Sen, in his book *Development as Freedom*, described political freedom as a roadmap to economic development. "Political freedoms help to promote economic security" (Sen, 1999b, p.11). Hornset and de Soysa (2022) used OLS regression to inspect how women's empowerment through the political sector is correlated with human capital development. The findings show that women's empowerment induces democracy and good authority. As a measure of human capital, the study takes human development, health, education, and child mortality rate as proxy variables. They follow Sundström et al.'s (2017) Varieties of Democracy's (VDEM) indicators on women empowerment. Endogenous growth theory is discussed to compare the study's outcome of human capital productivity. The study evaluates the concept of women's political empowerment and economic development, and the results show that women's empowerment is positively and significantly correlated with access to education. Political empowerment is considered a proxy variable for women's empowerment, ensuring equal access to education and health. So mainly, political empowerment induces human capital growth for women. Becker's unitary model infers that woman can make the same decision as men in the political sector of a nation (Mikkola and Miles, 2007). According to Kabeer (2005), the number of seats women hold in the national parliament is one of the crucial indicators of gender equality and women empowerment. She retaliated that since Bangladeshi women comprises half of the total population, they should be permitted at least half of the seats in the parliament. Again, specific qualifications are needed, which are required to be part of the parliament. Additinally, women face constraints in fulfilling these requirements. Studies by Kabeer (2005) also show that women in local government in India are more active in solving rural women's problems.

In South Asia, preferences for male child are culturally inherited; gender biases and high female mortality reflects the phenomenon that Amartya Sen called "missing women". In India, improvement in life expectancy is reducing the gender-biased mortality rate among adults.

Haque et al. (2021) assess the correlation between fertility rate and women empowerment. The study findings show that a lower fertility rate can induce women's empowerment. Tisdell and Hossain (2003) extensively studied the correlation between fertility and the female labour force, which results in unidirectional causality from the female labour force to fertility. Women's employment raises their autonomy and status and boosts control over decision-making power. They achieve financial independence, hence, an increase in self-esteem and lower fertility rate. The study found long-run relationships and causal links between fertility rate and female labour force using macro-level data in Bangladesh.

Application of dynamic time series model of vector error correction was applied, and bivariate model indicated unidirectional causal links between female labour force and fertility rate. The study also found that per capita GDP and infant mortality Granger causes fertility and female labour force. Anik et al. (2021) investigated the relationship between maternal health care and women empowerment among reproductive age from 15 to 49 in Bangladesh. The study with survey analysis found that women with a lower level of empowerment are less likely to take antenatal and postnatal care than higher empowered women. Also, Yaya et al. (2019) showed that in 34 sub-Saharan African countries, child marriage contributes to the fertility rate as women who were married as a child tend to have more children compared to women who married after the age of 18. According to World Health Organization (WHO) report, the adolescent (age 15–19) fertility rate in Bangladesh was one of the highest in 2020 (Bhowmik et al., 2021). So, the fertility rate strongly correlates with child marriage, mostly in developing countries. A lower fertility rate thus indicates women's family decision power and a lower rate of child marriage.

7.3 THEORETICAL APPROACHES

Hossain and Tisdell (2005) assessed different theoretic approaches to explain women's empowerment and gender inequality. They included universally acceptable leading theories like "The Incidence based criteria (included endowment and entitlement theory)", "The Unitary theory", and "The Bargaining theory" to explain women's status and empowerment (Sen, 1981; Schultz, 1990). Sen's theory talks about different entitlements like ownership entitlement, exchange entitlement, modes of production, and social security and employment entitlement. His theory explains that women are deprived of these entitlements and endowments from society. The Unitary theory suggests that the cumulative household decision-making process maximises the utility function of the family and is subject to collective budget constraints.

On the contrary, bargaining theories argue that there are no unified sets of family interests and that they face conflicts that threaten an individual's "power". Sen's theory is widely accepted because of its emphasis on an institutional approach to reducing inequality and poverty, and Becker's is more acceptable in the grassroots family decision-making process. The common ground of these theories is that they all require micro-level data on the female involvement in their family. On the other hand, macro-level data can shed light on women's empowerment by reducing the gender gap in societal, economic, and political contexts. Thus, inclinations in women's education and labour force participation can signify women's empowerment (Hossain and Tisdell, 2005).

7.3.1 Economic Development through Women Empowerment

Esther Duflo, a Nobel laureate in economics, contends that empowering women may be a significant force for development growth (Duflo, 2012). A few periods earlier, Amartya

Sen, another Nobel laureate, stated that the "capability" approach to development—also known as the institutional environment that fosters "freedoms" and gives people "agency" to pursue their aspirations for success—is what drives development (Sen, 1999b). Sen emphasises how lack of autonomy among the poor, particularly women, slows development and harms economic growth. The argument made by Duflo (2012) that the link between empowerment and development is too tenuous to self-regulate and that ongoing policy commitment to equality is required to ensure gender equality.

7.3.2 Research Gap

It is hard to find an appropriate framework for gender-biased social analysis (Sen, 1987). The aim of empowerment and reducing the gender gap fluctuate with different cultural, social, political, and regional frameworks. Most of the research works do not extensively study the elements of gender disparity and women empowerment, making it difficult on international platforms to evaluate and be assured about women's empowerment (Osmani, 2007). One regional study could not work for the other regional survey. Macro-level studies are weak in determining the agency, indicators and elements of women empowerment and the gender gap. A surfeit of micro-level studies has addressed women's empowerment and women's status in Bangladesh (Kabeer, 2017; Heath, 2014). But a limited number of studies are based on macro-level data. Also, there are inadequate studies that take more than one indicator of women's status in macro level data analysis to measure economic development through women empowerment. So, the present study intends to highlight an overview of women's empowerment in Bangladesh by depicting three critical indicators of women's status: female labour force participation, female literacy, and fertility rate.

7.4 DATA AND METHODOLOGY

The study uses data from World Development Index (WDI) and United Nations Development Organization (UNDP) to empirically estimate the relationship between women's empowerment and the economic development of Bangladesh. To examine the relationship, the *augmented Dickey-Fuller test*, *ARDL* for bound testing, *vector autoregressive* (VAR) statistical model, and *Granger causality* test have been used in this chapter.

7.4.1 Data

The study uses data on Bangladesh from WDI and UNDP, mainly from 1980 to 2021, which are inclusive sources of the development index. The dataset offers country-level data on the indicators of women's empowerment: female literacy rate, female labour force participation, and fertility rate. Also, for dependent variables, per capita GDP, and HDI data are collected from WDI and UNDP (see Table 7.1 for variable definitions).

7.4.2 Indicators of Women Empowerment

Three leading indicators have been chosen to measure women's empowerment.

7.4.2.1 Female Labour Force

Female employability is one of the major components of ensuring women's societal empowerment. Heath (2014), Hussain (2016), and Salway et al. (2005) emphasised that labour market opportunities can improve the lives of the women. The female labour force participation rate is a perfect proxy variable for women's empowerment because there is theoretical and empirical evidence that opportunities for women in the labour market may reduce early marriage and childbearing and strengthen women's bargaining power inside the family (Jensen, 2012; Rahman and Rao, 2004). Study also shows that working mothers increase their child's educational investment, which improves their opportunity to get better jobs and contributes to the economy (Oster & Millett, 2010). In this research, the percentage rate of the female labour force participation rate is taken from WDI, and the natural log form of the variable is taken to generate a less skewed

TABLE 7.1
Components Used to Construct a Quantitative Study

Independent Variable of Bangladesh (1980–2021)	Definition Given by World Development Index (WDI)
Female labour force participation (lnlf)	The percentage of adults (15 and older) who are economically active, or all those who work for a certain length of time to generate products and services.
Female literacy rate (lnltr)	The percentage of adults (15 and older) who can read and write well and comprehend a short, basic statement.
Fertility rate (lnfl)	A total number of children a woman would have if she had children at the age-specific fertility rates in the year in concern and lived to the end of her reproductive years.
Dependent Variable of Bangladesh (1980–2021)	**Definition Given by WDI**
GDP (lnGDP) (GDP per capita (constant 2015 US$))	GDP is estimated as the sum of the gross value added by all resident producers, plus any relevant product taxes, less any subsidies not represented in the price of the goods. Data are shown in current US dollars.
	Definition Given by United Nations Development Programme (UNDP)
Human development index (HDI/lnhdi)	HDI measures three fundamental elements of human development. It consists of life expectancy, an indicator of a long and healthy life. A person's (25 and older) average years spent in school is used to assess knowledge level. Expected years of schooling for school-age children measure access to learning. Gross national income (GNI) per capita (2017 constant foreign dollar) is converted using purchasing power parity (PPP) conversion rates as a measure of standard of life.

Note: lnlf, log(female labour force participation rate); lnltr, log(female literacy rate), lnfl, log(fertility rate); lnGDP, log(GDP per capita, constant 2015 USD); lnhdi, log(Human Development Index).

distribution and linearity in the model. Data only for adult women (15 years and above) who are economically active is taken for this research. The scatterplot shows a positive relationship between the independent variable female labour force participation and the dependent variables GDP per capita and HDI (Figures 7.1 and 7.2).

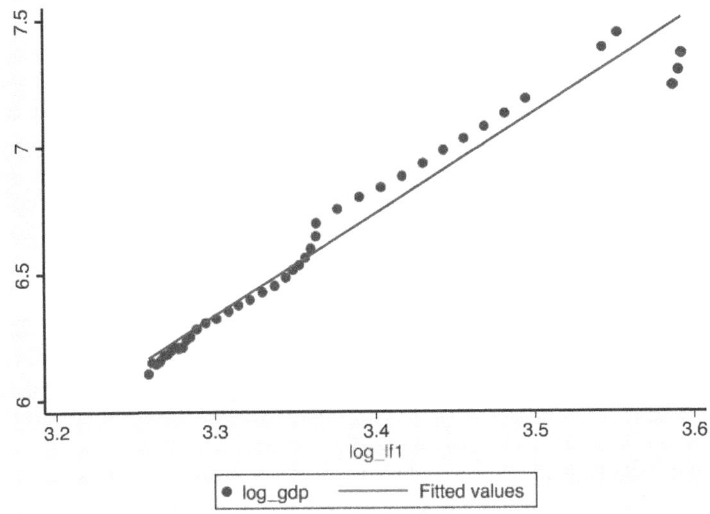

FIGURE 7.1 Scatterplot of ln_gdp and ln_lf1.

Note: ln_gdp: log(GDP per capita); ln_lf1: log(female labour force participation rate).

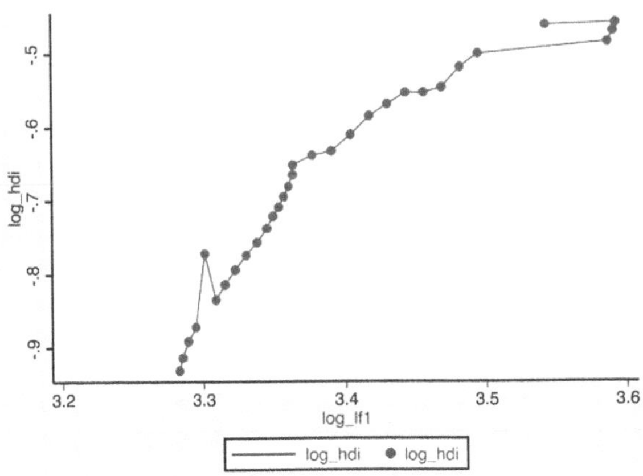

FIGURE 7.2 Scatterplot of ln_lf1 and ln_hdi.

ln_lf1: Natural logarithm of female labour force participation rate (age 15+); ln_hdi: Natural logarithm of the Human Development Index.

7.4.2.2 Fertility Rate

Women's desired fertility rate can be achieved by empowering them. So, the fertility rate reflects women's social independence and bargaining power within the family.

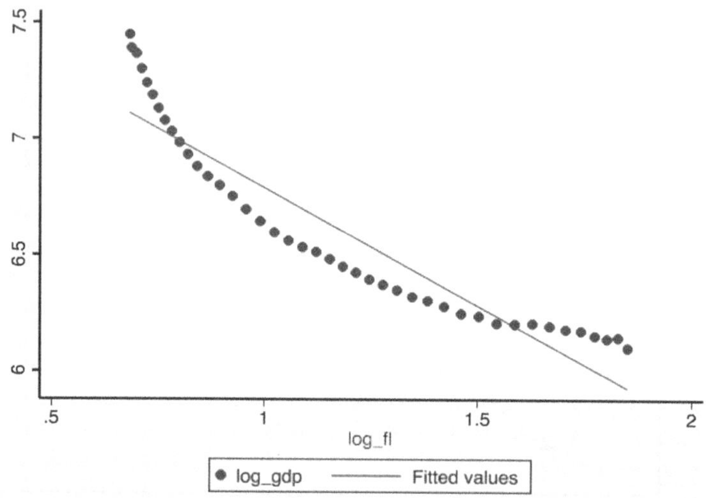

FIGURE 7.3 Scatterplot of ln_gdp and ln_fl.

ln_gdp: Natural logarithm of Gross Domestic Product per capita (constant 2015 US dollars); ln_fl: Natural logarithm of fertility rate (total births per woman).

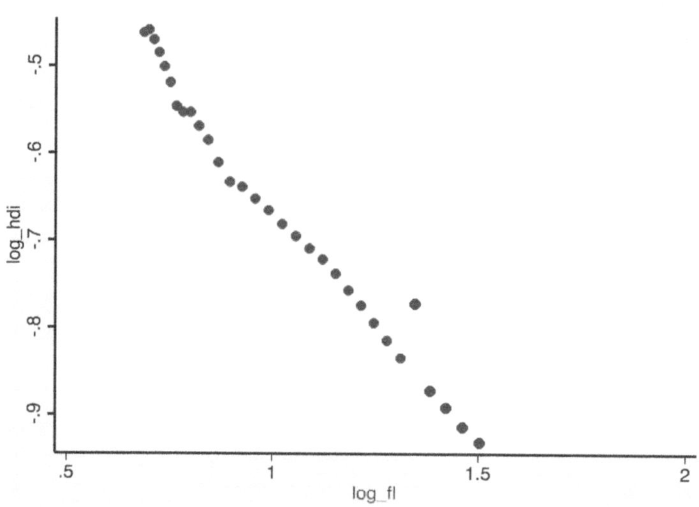

FIGURE 7.4 Scatterplot of ln_hdi and ln_fl.

ln_hdi: Natural logarithm of Human Development Index; ln_fl: Natural logarithm of fertility rate (total births per woman).

Also, child marriage contributes to the fertility rates, so lower fertility rate indicate lower child marriage (Yaya et al., 2019). In this research, total births per woman have been taken from WDI and transformed into a natural log form to generate less skewed distribution and linearity in the series. The scatter plot shows a negative relationship between the independent variable fertility rate and the dependent variables per capita GDP and HDI (Figures 7.3 and 7.4).

7.4.2.3 Female Literacy

The female literacy rate is another imperative variable to establish women's ability to question and change their inferior social standards. It also helps them access knowledge and information (Kabeer, 2005). Studies found that female literacy is highly correlated with the female labour force, which improves women's status in Bangladesh (Hossain and Tisdell, 2003; Aslam, 2014). In this chapter, the percentage rate of adult female literacy (15 years of age and above) has been taken from WDI as a proxy variable of women's status in Bangladesh. The data has been transformed into a natural log form for maintaining linearity and less skewed distribution in the model. The scatterplot represents a positive relationship between the variable female literacy rate and the dependent variables GDP per capita and HDI (Figures 7.5 and 7.6).

7.4.3 Model

Empirical analysis mostly depends on secondary quantitative data and multivariate regression analysis (Hossain and Tisdell, 2005). This study is based on a multivariate

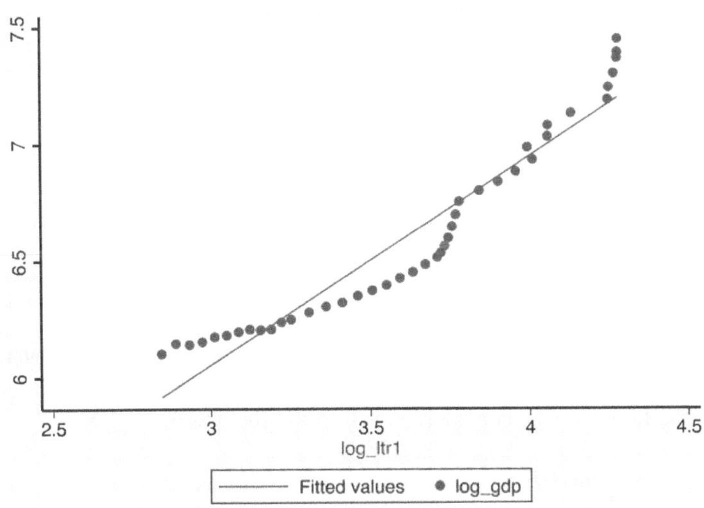

FIGURE 7.5 Scatterplot of ln_gdp and ln_ltr1.

ln_gdp:Natural logarithm of Gross Domestic Product per capita (constant 2015 US dollars); ln_ltr1: Natural logarithm of female literacy rate (age 15 and above).

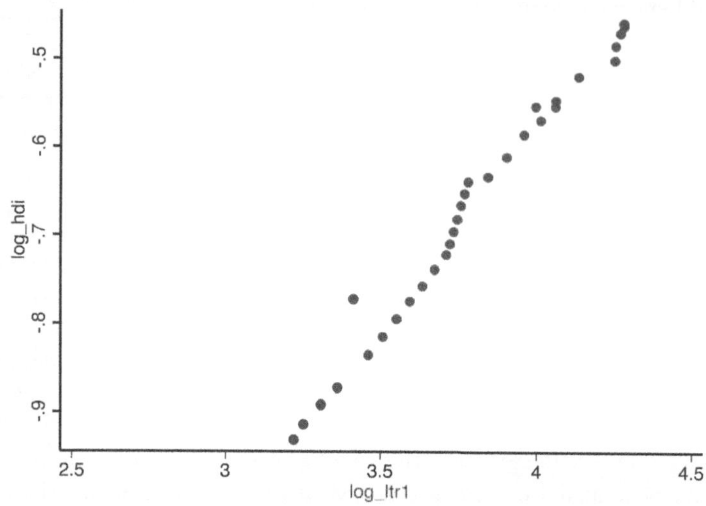

FIGURE 7.6 Scatterplot of ln_hdi and ln_Itr1.

ln_lf, Natural logarithm of female labour force participation rate (age 15+); ln_hdi, Natural logarithm of the Human Development Index.

time dimensional model using three main independent variables and two dependent variables taken from WDI and UNDP. The following model will be estimated:

$$\begin{aligned} \ln GDP = \alpha_0 + \beta_1 \ln(Female\ Labor\ Force\ (lf))_t \\ + \beta_2 \ln(Female\ Literacy(Itr))t \\ + \beta_3 \ln(Fertility\ Rate(fl)) + et \end{aligned} \tag{7.1}$$

$$GDP = f(Women\ Empowerment) \ldots \ldots \ldots \tag{7.2}$$

$$\begin{aligned} \ln HDI = \alpha_0 + \beta_1 \ln(Female\ Labor\ Force\ (lf))_t \\ + \beta_2 \ln(Female\ Literacy(Itr))_t \\ + \beta_3 \ln(Fertility\ Rate(fl))t + et \end{aligned} \tag{7.3}$$

$$HDI = f(Women\ Empowerment) \ldots \ldots \ldots \tag{7.4}$$

where β_1, β_2, and β_3, are the unknown population coefficients to be estimated through the data analysis, α_0 is the constant and e_t is the unobserved random error term.

A simple model can be considered to illustrate the ARDL model:

$$lnGDPt = \alpha + \beta\ lnflt + \delta lnlft + \gamma\ lnltrt + et \ldots . . \tag{7.5}$$

$$lnHDIt = \alpha + \beta\ lnflt + \delta lnlft + \gamma\ lnltrt + et \ldots . . \tag{7.6}$$

As in this research, there is cointegration among dependent and independent variables, so to test the long-run cointegrating relationship among variables, the error correction part of ARDL model is,

$$\Delta lnGDPt = \alpha 0 + \sum_{i=1}^{p} \beta i\ \Delta lnGDPt{-}i + \sum_{i=1}^{p} \delta i \Delta lnflt{-}i + \sum^{p} \gamma i \Delta lnlft{-}i + \sum_{i=1}^{p} \varepsilon i\ \Delta lnltrt{-}i$$
$$+ \lambda 1 lnGDPt{-}1 + \lambda 2 lnflt{-}1 + \lambda 3 lnlft{-}1 + \lambda 4 lnltrt{-}1 + ut \ldots \quad (7.7)$$

$$\Delta lnHDIt = \alpha 0 + \sum^{p} \beta i\ \Delta lnHDIt{-}i + \sum^{p} \delta i \Delta lnflt{-}i + \sum^{p} \gamma i \Delta lnlft{-}i + \sum^{p} \varepsilon i\ \Delta lnltrt{-}i$$
$$+ \lambda 1 lnHDIt{-}1 + \lambda 2 lnflt{-}1 + \lambda 3 lnlft{-}1 + \lambda 4 lnltrt{-}1 + ut \ldots. \quad (7.8)$$
$$i=1 \qquad i=1 \qquad i=1$$

$$\Delta\ \mathbf{lnGDPt} = \sigma + \sum_{i=1}^{n} ai\boldsymbol{\Delta lnGDPt{-}i} + \sum_{j=1}^{n} \beta j\ \Delta lnflt{-}j + \sum_{j=1}^{n} \gamma m\ \Delta lnlft{-}m$$
$$+ \sum_{k=1}^{n} \delta k\ \Delta logltrt{-}k{+}u1t \ldots\ldots \quad (7.9)$$

$$\boldsymbol{\Delta lnflt} = d + \sum_{i=1}^{n} \alpha i \Delta lnGDPt{-}i + \sum_{\boldsymbol{j=1}}^{\boldsymbol{n}} Qj\ \boldsymbol{\Delta lnflt{-}j} + \sum_{j=1}^{n} \gamma m\ \Delta lnlft{-}m$$
$$+ \sum_{k=1} \delta k\ \Delta lnltrt{-}k{+}u2t \ldots\ldots \quad (7.10)$$

$$\boldsymbol{\Delta lnlft} = b + \sum_{i=1n}^{n} \alpha i \Delta lnGDPt{-}i + \sum_{j=1}^{n} \beta j\ \Delta lnflt{-}j + \sum_{\boldsymbol{j=1}}^{\boldsymbol{n}} m\ \boldsymbol{\Delta lnlft{-}m}$$
$$+ \sum_{k=1} \delta k\ \Delta logltrt{-}k{+}u3t \ldots\ \ldots. \quad (7.11)$$

$$\Delta lnltrt = \rho + \sum_{i=1\boldsymbol{n}}^{n} \alpha i \Delta lnGDPt{-}i + \sum_{j=1}^{n} \beta j\ \Delta lnflt{-}j + \sum_{j=1}^{n} \gamma m\ \Delta lnlft{-}m$$
$$+ \sum_{\boldsymbol{k=1}} \boldsymbol{k}\ \boldsymbol{\Delta lnltrt{-}k}{+}u4t \ldots\ldots \quad (7.12)$$

Here, n = the optimal lag length, α_i, β_j, γ_m, and δk are the short coefficients of the above-mentioned model. Eq. (7.9) shows that the current value of GDP is related its past values and the past values of female labour force (lnlf), literacy rate (lnltr) and fertility rate (lnfl).

Similarly, Eqs. (7.10–7.12) offers that *lnfl, lnlf, and lnltr* are related to the past values of themselves, respectively and that of GDP. The null hypothesis in Eq. (7.9) is, $\beta_j = 0$, $\gamma_m = 0$ *or* $\delta k = 0$ which means, "lnfl, lnlf or lnltr does not Granger cause

GDP". Similarly, the null hypothesis in Eqs. (7.10–7.12) is $\alpha i = 0$, and states "GDP does not Granger cause *lnfl, lnlf, lnltr*" (Das, 2019).

The model of Granger causality of this research when the dependent variable is HDI is as follows,

$$\Delta \mathbf{lnHDIt} = \sigma + \sum_{i=1}^{n} ai\Delta \mathbf{lnHDIt{-}i} + \sum_{j=1}^{n} \beta j\ \Delta lnflt{-}j + \sum_{j=1}^{n} \gamma m\ \Delta lnlft{-}m + \sum_{k=1}^{n} \delta k\ \Delta lnltrt{-}k{+}u1t \ldots\ldots \quad (7.13)$$

$$\Delta \mathbf{lnflt} = d + \sum_{i=1}^{n} \alpha i\Delta lnHDIt{-}i + \sum_{\mathbf{j=1}}^{\mathbf{n}} Q\mathbf{j}\ \Delta \mathbf{lnflt{-}j} + \sum_{\mathbf{j=1}}^{n} \gamma m\ \Delta lnlft{-}m + \sum_{k=1}^{n} \delta k\ \Delta logltrt{-}k{+}u2t \ldots\ldots \quad (7.14)$$

$$\Delta \mathbf{lnlft} = b + \sum_{i=1}^{n} \alpha i\Delta lnHDIt{-}i + \sum_{j=1}^{n} \beta j\ \Delta lnflt{-}j + \sum_{\mathbf{j=1}}^{\mathbf{n}} \mathbf{m}\ \Delta \mathbf{lnlft{-}m} + \sum_{k=1}^{n} \delta k\ \Delta lnltrt{-}k{+}u3t \ldots\ \ldots. \quad (7.15)$$

$$\Delta lnltrt = \rho + \sum_{i=1}^{n} \alpha i\Delta lnHDIt{-}i + \sum_{j=1}^{n} \beta j\ \Delta lnflt{-}j + \sum_{j=1}^{n} \gamma m\ \Delta lnlft{-}m + \sum_{\mathbf{k=1}}^{\mathbf{n}} \mathbf{k}\ \Delta \mathbf{lnltrt{-}k}{+}u4t \ldots\ldots \quad (7.16)$$

Eq. (7.13) shows that the current value of HDI is related to its past values and the past values of female labour force (lnlf), female literacy rate (lnltr) and fertility rate (lnfl). Similarly, Eq. (7.14–7.16) postulates that *lnfl, lnlf, and lnltr* is related to the past values of themselves, respectively and that of HDI. The null hypothesis in Eq. (7.13) is, $\beta_j = 0$, $\gamma_m = 0$ *or* $\delta_k = 0$ which means, "lnfl, lnlf, or lnltr does not Granger cause HDI". Similarly, the null hypothesis in Eq. (7.14–7.16) is $\alpha_i = 0$ and states, "GDP does not Granger cause *lnfl, lnlf, lnltr*". The selection of rejection or non-rejection area can be selected based on the F-statistics from Error Correction Model (ECM) test results.

7.5 RESULTS

In this section, results are described and deliberated in the order in which they were introduced in part of the data and methodology (Section 7.4). In brief, subsection 7.5.1 shows the summary statistics of the sample variables are presented in Table 7.2, subsection 7.5.1.1 emphasises on the properties of the data series and demonstrates unit root testing for dependent and independent variables. Subsection 7.5.3 delves into estimating cointegration tests for assessing the long run and the short run relationship between the dependent and independent variables in Figure 7.7, we can see that residuals are normally distributed for depended variables, HDI and GDP from top to the next one respectively. Finally, subsection 7.5.1.1 exhibits the assessment of the direction of the long-run causal relationship.

7.5.1 Summary Statistics of the Sample

The descriptive statistics of all variables are presented in Table 7.2.

TABLE 7.2
Summary Statistics of the Sample

Variable	Obs	Mean	Std. Dev.	Min	Max	Variance	Skewness	Kurtosis
fl	42	3.496738	1.374706	1.979	6.359	1.889816	.6799413	2.177671
lf	42	29.17509	3.059562	26.0069	36.311	9.360922	1.044161	3.031859
ltr	42	40.87682	17.40836	17.1849	72.0611	303.0511	.4473114	2.049639
GDP	42	808.2009	366.3757	447.694	1715.35	134231.2	1.043284	2.908648
HDI	31	.5160645	.0719884	.394	.632	.0051823	.0189205	1.893596

Author's calculation is based on WDI and UNDP data.
Variable definition: fl: fertility rate; lf: female labour force; ltr: female literacy rate.

TABLE 7.3
ARDL–ECM, Long-Run and Short-Run Cointegration Estimation, Dependent Variable: GDP

Long-Run Estimated Coefficients

Variable	Coefficient	Standard Error	P-Value
lnlfl (female labour force)	1.900708***	.6346115	0.006
lnfl (fertility rate)	–1.231641***	.4139101	0.006
lnltr1 (female literacy)	–0.5550882	.4312153	0.209

Short-Run Estimated Coefficient

Variable	Coefficient	Standard Error	P-Value
lnlfl	0.2719972**	.097549	0.010
lnfl	1.87282	4.81202	0.700
lnltr1	-0.0794348	.0558421	0.166

Akaike Information Criterion (AIC) is used to select optimum lag length; ***, **, * indicate statistical significance at 1%, 5% and 10% levels, respectively.
ARDL $R^2 = 0.9997$
F-Statistics = 0.000
ADJ term, ECT = -0.1431031, P-value = 0.008, Constant = 0. 6197603

TABLE 7.4
ARDL–ECM, Long-Run, and Short-Run Cointegration Estimation, Dependent Variable: HDI

Long-Run Estimated Coefficients			
Variable	**Coefficient**	**Standard Error**	**P-Value**
lnlfl (female labour force)	-0.165331	0.2506939	0.515
lnfl (fertility rate)	-0.46964***	0.14396	0.003
lnltrl (female literacy)	0.101599	0.1861351	0.590

Short-Run Estimated Coefficient			
Variable	**Coefficient**	**Standard Error**	**P-value**
lnlfl	0.2030886	.128095	0.125
lnltrl	0.14199*	0.08189	0.095
lnfl	-0.3489942***	0.0910625	0.001

To select optimum lag length, Akaike Information Criterion (AIC) is used.
***, **, * indicate statistical significance at 1%, 5% and 10% levels, respectively.
ARDL, $R^2 = 0.9886$
F-Statistics = 0.000
ADJ term, P-value = 0.004, ECT = -0.53439, Constant = −1.596442

7.5.1.1 Cointegration Test Results When the Dependent Variables Are Individually GDP and HDI

Results of long-run and short-run estimates are as follows according to the Eqs. (7.7) and (7.9). See Figure 7.7 and Tables 7.3 and 7.4.

Here, we have cointegration relationships in the models through ARDL bound testing, so we can test vector autoregressive model (VAR) with variables at the level to obtain long-run causality between the cointegrating variables in the model.

As described in subsection 7.5.1.2, Granger causality shows a pair-wise relationship in the long run, which can be a one-way or two-way relationship between included variables. This Granger causality test validates the cointegration relationship and shows us the direction of the long run relationships. So, ARDL-ECM models and VAR testing at the level of the variables show us the long-run Granger causality.

In a nutshell, the female literacy rate and per capita GDP of Bangladesh has bidirectional Granger causality (Tables 7.5, 7.6, and 7.7). The VAR testing and Granger causality Wald testing has been estimated on the variables at their level without controls for lags and trend that is why it is showing a causal relationship even though there is no significant cointegration relationship in female literacy and per capita GDP.

Also, the fertility rate has a bidirectional Granger causality with the per capita GDP of the country. And per capita GDP causes female labour force, which is a unidirectional Granger causality. So, the Granger causality Wald test shows the pair-wise long-run causal relationship between the fertility rate and GDP per capita and female labour force and per capita GDP. Moreover, fertility rate and female labour force have

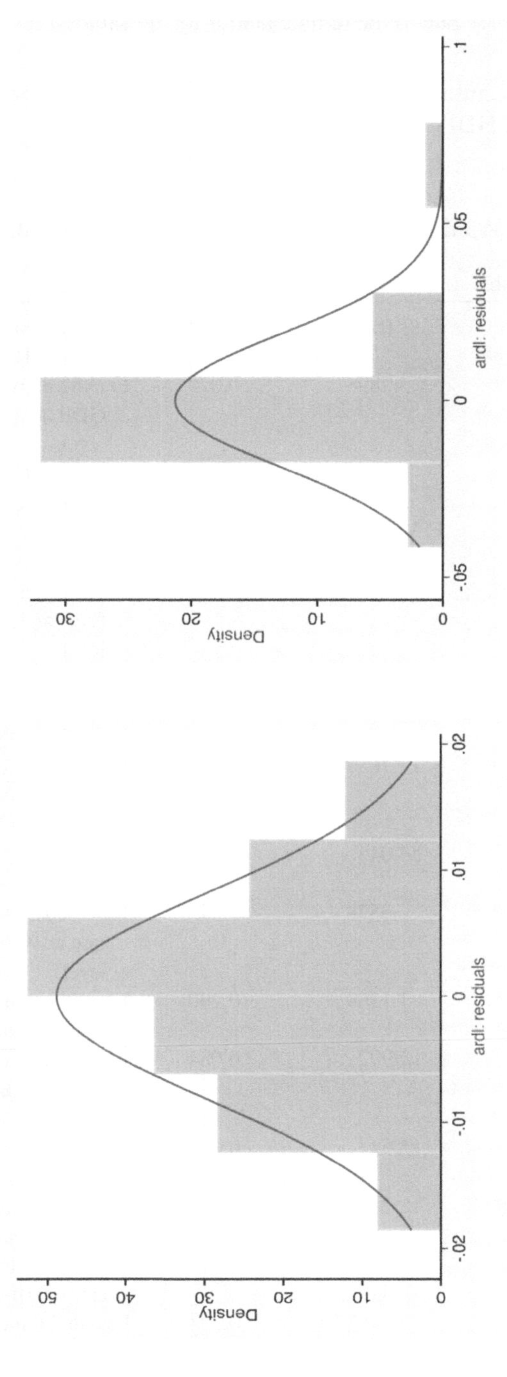

FIGURE 7.7 Residuals are normally distributed where depended variables are HDI and GDP from top to the next one respectively.

a unidirectional Granger causality, and the causality runs from fertility rate to female labour force. Also, results show that there is unidirectional Granger causality that runs from the female literacy rate to the female labour force (Tables 7.5, 7.6, and 7.7).

7.5.1.2 Granger Causality When the Depended Variables Are Individually GDP and HDI

TABLE 7.5
Granger Causality Wald Test Result, Dependent Variable Is GDP in the Model

Equation	Variables	Chi2	Prob > Chi2	Decision
ln_gdp	ln_lf1	.01189	0.913	Female labour force does not cause GDP
ln_gdp	ln_ltr1*	3.2898	0.070	Female literacy rate causes GDP at 10% significant level
ln_gdp	ln_fl***	7.665	0.006	Fertility rate causes GDP at 1% significant level
ln_gdp	ALL	30.829	0.000	
ln_lf1	ln_gdp***	42.645	0.000	GDP causes female labour force at 1% significant level
ln_lf1	ln_ltr1***	16.003	0.000	Female literacy rate causes female labour force at 1% significant level
ln_lf1	ln_fl***	17.76	0.000	Fertility rate causes female labour force at 1% significant level
ln_lf1	ALL	54.045	0.000	
ln_ltr1	ln_gdp*	2.9578	0.085	GDP causes female literacy rate at 10% significant level
ln_ltr1	ln_lf1	1.1209	0.290	Female labour force does not cause female literacy rate
ln_ltr1	ln_fl*	2.992	0.084	Fertility rate causes literacy rate at 10% significant level
ln_ltr1	ALL	15.543	0.001	
ln_fl	ln_gdp***	18.281	0.000	GDP causes fertility rate at 1% significant level
ln_fl	ln_lf1	.83787	0.360	Female labour force does not cause fertility rate
ln_fl	ln_ltr1	.14577	0.703	Female literacy rate does not cause fertility rate
ln_fl	ALL	95.811	0.000	

To select optimum lag length, Akaike Information Criterion (AIC) is used.
***, **, * indicate statistical significance at 1%, 5% and 10% levels, respectively.

7.5.1.3 Granger Causality When the Dependent Variable Is GDP

TABLE 7.6
Direction of Granger Causality Test, Dependent Variable in the Model: GDP

Variable	Causality Direction in the Long Run
Female literacy rate causes GDP	Bidirectional Granger causality
Fertility rate causes GDP	Bidirectional Granger causality
GDP causes female labour force	Unidirectional Granger causality
Fertility rate causes female literacy rate	Unidirectional Granger causality
Female literacy causes female labour force	Unidirectional Granger causality
Fertility rate causes female labour force	Unidirectional Granger causality

7.5.1.4 Granger Causality When the Dependent Variable Is HDI

TABLE 7.7
Granger Causality Wald Test Result, Dependent Variable HDI in the Model

Equation	Variables	Chi2	Prob > Chi2	Decision
ln_hdi	ln_lf1*	5.3847	0.068	Female labour force causes HDI at 10% significant level
ln_hdi	ln_ltr1	3.7676	0.152	Female literacy rate does not cause HDI
ln_hdi	ln_fl***	21.62	0.000	Fertility rate causes HDI at 1% significant level
ln_hdi	ALL	52.435	0.000	
ln_lf1	ln_hdi	.02312	0.989	HDI does not cause female labour force
ln_lf1	ln_ltr1***	14.091	0.001	Female literacy causes female labour force at 1% significant level
ln_lf1	ln_fl	.31066	0.856	Fertility rate does not cause female labour force
ln_lf1	ALL	31.397	0.000	
ln_ltr1	ln_hdi	.86165	0.650	HDI does not cause female literacy rate
ln_ltr1	ln_lf1	.83017	0.660	Female labour force does not cause female literacy
ln_ltr1	ln_fl***	20.824	0.000	Fertility rate causes literacy rate at 1% significant level
ln_ltr1	ALL	25.625	0.000	
ln_fl	ln_hdi	1.7485	0.417	HDI does not cause fertility rate
ln_fl	ln_lf1***	15.21	0.000	Female labour causes fertility rate at 1% significant level
ln_fl	ln_ltr1***	65.867	0.000	Female literacy causes fertility rate at 5% significant leve
ln_fl	ALL	76.863	0.000	

To select optimum lag length, Akaike Information Criterion (AIC) is used.
***, **, * indicate statistical significance at 1%, 5% and 10% levels, respectively.

7.6 DISCUSSION

There is bidirectional causality between women's empowerment and economic development (Duflo, 2012). This chapter will concisely discuss both sides of the causal direction of women's empowerment and economic development relationship in light of the eminent article "Women Empowerment and Economic Development" by Esther Duflo (2012).

Besides, for the completion of this chapter, the book *Development as Freedom* by Amartya Sen (1999b) was predominantly consulted to establish a connection between the actual exposure and theoretical and empirical findings of this chapter.

Additionally, women and girls endure a wide range of deprivation and discrimination due to their gender and its relationship to poverty in Bangladesh (Hossain, 2020). According to the global gender gap index rankings, Bangladesh ranked 65th in 156 countries (World Economic Forum, 2025). While Bangladesh scored 7th in political empowerment, it lagged in other sub-indices, placing 147th in economic participation and opportunity, 121st in education, and 134th in health (World Economic Forum, 2025). Similarly, gender equality is acknowledged in the 2000 Millennium Declaration as being essential to accomplishing all the Millennium Development Goals (MDGs). The World Bank claims that increasing gender equality encourages primary education for everyone, lowers infant mortality, and improves mother health. Indirectly, it affects family well-being through the impact of women's enhanced household decision-making, and it directly improves poverty reduction through women's higher labour force participation, productivity, and incomes (World Bank, 2023). So, it is essential to comprehend what Bangladesh has accomplished as the world prepares to reach the Sustainable Development Goals (SDGs). In the 1990s and 2000s, Bangladesh achieved startlingly fast and continuous growth on poverty reduction and gender equality (Hossain, 2020).

Sen (1999b) mentioned that women's well-being is significantly related to their ability to earn independently. From 2019 to 2021, there has been a prominent gender gap in the labour force in Bangladesh. We can see that in 2016 and 2021, the female labour force participation for 15+ age is respectively 33% and 35% approximately, whereas the rate for males in the same year is respectively 78.7% and 78.7%. The adult women's participation rate in the labour market rose by 10% from 2003 to 2016 but still, the female labour force participation rate is now only 35% (age 15+) (World Bank, 2021).

Moreover, the most considerable unemployment rates are among rural females aged 15 to 34. The continuance of child marriage remains a significant obstacle even though married women now participate in the labour field in Bangladesh at substantially higher percentages than they were ten years ago. Married women are now more prevalent in the job field than in the past is in fact a positive development.

However, the World Bank data found that many women still require their husbands' or in- laws' approval to work outside the house after marriage. Despite a drop in sex discrimination across many industries, women still seem to have remarkably few opportunities for employment in a wide range of occupations in Bangladesh (Solotaroff et al., 2019).

According to statistics from the World Bank Labour Force Survey (LFS), women working in industries with female dominance fell by more than two-thirds. But women's bargaining power has not increased as much as men's outside of agriculture, that is, in the industry and service sector. Boys tend to gravitate to science subjects which provide skills for the job market, whereas females predominantly do not.

As a result, there is still significant occupational sex segregation in Bangladesh. Gender-based occupational segregation is more prevalent in rural than urban settings; middle-class urban women have access to a wider range of jobs than urban poor and rural women (Solotaroff et al., 2019). Women are paid less for doing the same job as males in low- and middle-income countries. Also, despite participating in the labour force, they are more likely to remain poor (Duflo, 2012). According to Duflo (2012), women in developing countries spend twice time than man in households, five times than men in childcare and half time than men in working in the economy. This study also mentioned that 21 studied countries have unequal property rights for males and females, which is also factual for Bangladesh. Women have fewer labour market opportunities than males in developing countries, which can be the main reason for unequal treatment between males and females in households. It has been shown in several studies that parents who believe that return to education will be higher are more likely to invest in their child's education. In developing countries, parents have more ambition for their sons than daughters (Duflo, 2012). Also, Qian (2008) shows that the number of missing women, which is strikingly high in China, decreased in the tea-producing regions compared to the other areas.

Through a randomized controlled trial (RCT) program, introducing Business Process Outsourcing Centers (BPOs) in a village in India resulted in a 5% increase in girls' enrolment in school as parents are taking better care of them. This evidence complements the findings of this chapter as it shows that in the long run, a 1% increase in female labour force participation will increase the per capita GDP by 1.900708% at a 1% significant level in Bangladesh. And in the short run, a 1% increase in female labour force participation will increase 0.27199% per capita GDP at a 5% significant level in Bangladesh (Table 7.3). Also, causality tests from this study show that in the long run, the female labour force causes HDI, and GDP causes the female labour force, which is both unidirectional Granger causality (Tables 7.8 and 7.5, respectively). All this evidence shows that increasing women's opportunities in the labour market can reduce gender discrimination in households, moving them towards gender equality and economic development (Duflo, 2012).

Bangladesh's educational system consists of primary (grades 1 through 5), secondary (grades 6 through 12), and higher secondary schools. Likewise, several private schools offer English-medium instruction. They provide courses at the "O" and "A" levels. Primary school age is 6 years, whereas lower secondary, secondary, and upper secondary are, respectively, age 11 to 13, 14 to 15, and 16 to 17. We can see that there is a major gap between female and male literacy in Bangladesh.

However, the gap has been rapidly closing recently. In 2021, the adult female (age 15+) literacy rate was 72.06%, whereas the adult male literacy rate was 77.8%. Also, from 2010 to 2015, girl's enrolment in primary school increased from 112.4% to 113.4%, whereas for boys, it increased from 103.2% to 105%, but the pass rate of boys in primary education was higher than a girl in this period (Shilpi et al., 2017). The dropout rate was also higher for girls than boys from 2010 to 2015, and this dropout rate from primary school is still higher for girls than boys. The International Labour Organization (2015) from a child labour survey shows that most of the reason for the dropout of girls was financial problems, and for boys, it was due to disinterest of the parents and child in 2013. Also, the dropout rate is highest in grade 4 for both girls and boys. In secondary education, the participation rate of girls was more than the boys from 2010 to 2015. In 2010, the total enrolment rate of girls was 62.62% and for boys it was 49.595% which increased to 77.88% and 67.75%, respectively, in 2015. Even though the participation rate of girls in secondary education is higher than boys, the completion rate at the secondary level was lower for girls than boys. The dropout rate for girls from the secondary level was 46% in 2015, whereas the dropout rate was 34% for boys that year. Again, the child labour survey by the International Labour Organization (2015) shows that this large number of dropouts of girls from secondary school is primarily due to domestic chores. For boys, it was to support the family income. The educational gap is much more significant in upper secondary education, where male participation is consistently above 50%, and female participation is below 50% from 2010 to 2015. Also, female enrolment sharply declined in 2015 from 52% to 36% (International Labour Organization, 2015). Even female education completion rate was lower than males in 2015. According to Duflo (2012), in low and middle-income countries, the secondary education attainment for girls was 34%, whereas it was 41% for boys in 2010. Studies show that the correlation between a mother's earnings and education to a child's well-being is more robust than a father's earnings and education. But correlation can be misleading. If women are the main caretaker for their children, then more education for women will have more spill-over effect than giving more education to men (Duflo, 2012). This chapter shows no significant relationship between women's literacy and GDP growth rate, which supports Duflo's educational arguments (2012). On the contrary, it has a relationship with HDI as a 1% increase in female literacy is associated with a .14199% increase in HDI in Bangladesh at 10% significant level (Table 7.4). But according to Duflo, the concept that education is more important for female than male for better children outcome and economic growth need to be still revised as education for both male and female is equally indispensable (Duflo, 2012).

7.6.1 Nexus between Economic Development and Women Empowerment

Women empowerment and economic development has a bidirectional causal relationship (Duflo, 2012). Improving the capabilities and accessibility of women, particularly in education, labour force, health, and the political sector, will result in the country's economic development. In one direction, economic development will reduce gender inequality.

According to Sen, discrimination against women will hinder the development of a country. In the other direction, women's empowerment will accelerate economic development (Duflo, 2012). Duflo also mentioned that as the country develops, women's empowerment will follow a positive relationship with the development. So economic development reduces poverty and gender inequality by increasing available household resources. This also reduces the excess vulnerability of women. When a family cannot have enough money to feed everyone, then they excessively sacrifice the welfare of girls during a natural disaster (Rose, 1999). This also suggests that reducing poverty and enhancing resources for households can improve the welfare of women. So Rose (1999) shows that households can tackle natural disasters and reduce discrimination against women if the families have enough resources, so the mortality rate of girls in developing countries during droughts is less in the little wealthier households. Economic development reduces poverty and increases the welfare of families.

So, reducing the vulnerability of the poor with policy actions without only targeting women will result in economic development and improvement of their well-being (Duflo, 2012). So, will economic development be sufficient to reduce gender inequality and stimulate women's empowerment? The answer is no. There is evidence that only economic growth cannot overcome gender discrimination. Studies show that the availability of new technologies and increased household well-being raises gender discrimination.

Technological advancements reduced the cost of sex identification and abortion. As a result, sex-selective abortion rose in India in 2011 as parents wanted to avoid dowry, which they would otherwise have to pay during their daughter's marriage (Duflo, 2012).

This chapter shows that in the long run fertility rate has a bidirectional Granger causality with GDP per capita in Bangladesh. And GDP per capita has a unidirectional Granger causality with the female labour force where the causality runs from GDP to the female labour force (Table 7.7). The results also divulge that the adult female (15 and older) literacy rate has a positive significant correlation with HDI in Bangladesh. Also, the adult female (15 and older) labour force has a positive significant correlation with per capita GDP. And the fertility rate has a significant negative correlation with GDP and HDI. (Additionally, in Nepal reduction in the mortality rate of women due to childbirth causes economic development (Duflo, 2012). These observations are consistent with the findings of this paper as this paper shows that the fertility rate has a long-run negative relationship with GDP, as a 1% increase in the fertility rate will result in a 1.231641% decrease in the per capita GDP in Bangladesh at 1% significant level (Table 7.3) in the short run, a 1% increase in fertility rate will decrease the HDI of Bangladesh by 0.46964% on average at a 1% significant level (Table 7.4). Also, in the long run, a 1% increase in fertility rate will decrease the HDI of Bangladesh by 0.3489942% on average at a 1% significant level.

Moreover, the fertility rate shows a bidirectional granger causality with GDP, and It has unidirectional granger causality with female literacy, where the causality runs from fertility rate to female literacy, which means if the fertility rate reduces, then that will cause female literacy to rise (Table 7.7).

So, we can see the indicators of women empowerment have a causal and cointegration relationship with the GDP and HDI of Bangladesh. So, it can be opined that women's empowerment and economic development in Bangladesh are interrelated, which also supports the findings of previous literature mentioned in this Section 7.6 and Section 7.2.

7.7 POLICY RECOMMENDATION AND CONCLUSION

7.7.1 Policy Recommendation

There is a debate among policymakers about which direction to focus on, the causal relationship between economic development to women empowerment or women empowerment to economic development. Policymakers focus on one direction between these two relationships. Those who support the causal relationship from economic development to gender inequality emphasise creating conditions for economic growth without adopting specific programs for improving the condition of women.

On the contrary, others emphasise women's empowerment for achieving economic development.

Quite the reverse, Doepke and Tertilt (2009) suggest that institutions like the World Bank should aim for economic development programs such as educational policy enhancement rather than just seeking immediate policy changes for women's rights. They suggest that as a country develops, it will empower women. Moreover, poverty reduction with economic development can result in a decrease in gender inequality. Rose (1999) shows that in natural disasters like droughts in India, providing insurance against risk management will reduce the child mortality rate. Likewise, pension programs like in South Africa have evidence to improve the condition of women. So, in the early 1990s, old age pension programmes in South Africa reduced the witch killing in rural northern Province (Duflo, 2012). However, policies that only target women need to be rationalised as these policies would be formed at the expense of men. Duflo showed examples in this regard. She argued that any position women get in politics through quota is blocking the position for men. Also, within the most constrained budgets of the developing countries, giving scholarships for education just to girls is very expensive and comes at the direct expense of boys.

The money spent on only girls' education is not spent on other things like new teachers and deworming programs; also, programs like informing the parents about the return of education would otherwise benefit both the girls and boys. Yet conditional cash transfer programs targeting girls result in higher returns than boys in developing countries (Duflo, 2012).

So, benefit programs like "PROGRESA" in Mexico, insurance programs against risk management and conditional cash transfer programs in India and old age pension programmes in South Africa can be implemented in Bangladesh to reduce the gender gap in the country.

Also, from a political perspective, in India, in 1993, constitutional amendment made it compulsory that one-third of the rural council seat and rural presidencies seats be reserved for women (Duflo, 2012), which is uncommon in Bangladesh. In Bangladesh, one-third of Union Councils (or Union Parishads or Unions) seats can be reserved for women, which are now one forth according to the parliament act of 1997 in Bangladesh (laws of Bangladesh, 1997). Also, as mentioned earlier in the literature section, according to Kabeer half of the total population, women in Bangladesh should be permitted at least half of the seats in the parliament (Kabeer, 2005). Moreover, child marriage in Bangladesh contribute to the high fertility rate and one of the main obstacles to the empowerment and freedom of a woman. So, policy should be strict to defend these early marriages specially in poor population in the country. Therefore, the government of Bangladesh should implement policies which will increase female participation in the public sector to promote gender equality in the public sector.

Besides, opportunities for women outside their marriage will enhance the household's bargaining power. So, in case of divorce, if the women rights are not strong enough in the country and there are not enough labour market opportunities, this would negatively affect women's bargaining power in the family. So, according to Duflo, policies aiming to increase the welfare of divorce and women's approach to the labour market will enhance the status of women in the family. Added to that, property right is a big problem in Bangladesh.

Substantial property rights should be imposed to enhance the well-being of women in a household. Even a minor change in women's control over recourse in a home will increase their say in the family (Duflo, 2012). As Sen also said, property rights will make women more powerful in making family decisions (Sen, 1999b).

7.7.2 Conclusion

This study aims to answer the question posed at the beginning, "is there any cointegration and causal relationship between women empowerment and the economic development of Bangladesh?". Primarily, the findings discussed in this study underline the importance of women's empowerment in the economic development of Bangladesh. This study discusses the knowledge of women's empowerment, which is conceptualised as a multi-dimensional change process. Female literacy rate, female labour force participation, and fertility rate are the three key indicators to quantify this change process in women's lives.

These three key factors touch on many features of women's lives, including their self-worth, social standing, willingness to challenge their inferior status in society, bargaining power in the household and finally, to improve their capacity to work equally with men to contribute to the development of the country.

Also, in the discussion part (Section 7.6), this chapter demonstrated the supporting evidence for the findings of this study. The study findings show a significant cointegration and causal relationship between women's empowerment and economic development. More precisely, the three main proxy variables for women empowerment (female labour force participation, fertility rate and female literacy rate) show a significant relationship with the GDP and HDI of Bangladesh. Firstly, the results show that increasing female labour force participation will significantly increase the GDP and HDI of Bangladesh. Similarly, increasing the female literacy rate will increase the HDI in the country. But on the contrary female literacy has no significant cointegration relationship with the country's GDP.

Moreover, increasing the fertility rate will significantly decrease the country's HDI and GDP. These results are highly robust. Also, we see that there are supporting evidence for the findings of this chapter (Section 7.6), as studies show that mainly in developing countries, increasing female labour force participation and female literacy rate, and lowering fertility rate results in the higher development of a country (Duflo, 2012, Sen, 1999b).

Most females are under-represented in the labour force and literacy sector of Bangladesh. Also, there is a high number of adolescent marriages in the country, and low fertility rates indicate higher family decision power among women. Thus, increasing female participation in the labour force and literacy and reducing the fertility rate will result in the higher economic development of the county. Therefore, policies that will increase female participation in the labour market and educational sector and which will also encourage a lower fertility rate should be implemented by the government to promote women empowerment and gender inequality in Bangladesh.

So, the concluding part can be summarised with the result and evidence that women's empowerment and economic development are closely interconnected in Bangladesh. Women's empowerment will bring changes in decision-making with higher participation of women in the labour market and educational sector, as a result, the fertility rate will be lower, which will have a direct impact on the development of the country. Furthermore, as mentioned earlier, two important variables have been omitted from the study for the lack of data availability, so an interesting opportunity for future research would be to investigate the impact of female participation in the political sector and access to microfinance by women on women's empowerment and the economic development of Bangladesh.

BIBLIOGRAPHY

Alderman, H., Chappori, P.A., Haddad, L., Hoddinot, J. and Kanbur, R. 1995. Unitary vs collective models of households: is it time to shift the burden of proof? *World Bank Research Observer*, Vol. 10 No. 1, pp.1–19.

Alsop, R. and Bryan K. 2005. *Local Organizations in Decentralized Development: Their Functions and Performance in India*. The World Bank Group.

Al-Zayed, S. R., Talukdar, F., Jahan, F., Asaduzzaman, T. M., and Shams, F. 2018, December. Beyond gender parity: Actualization of benefits versus fallacy of promises – A case study on Bangladesh (World Bank Report No. 32012). World Bank. https://doi.org/10.1596/32012

Anderson, S. and Eswaran, M., 2009. What determines female autonomy? Evidence from Bangladesh. *Journal of Development Economics*, 90(2), pp.179–191.

Anik, A., Ghose, B. and Rahman, M., 2021. Relationship between maternal healthcare utilisation and empowerment among women in Bangladesh: evidence from a nationally representative cross-sectional study. *BMJ Open*, 11(8), pp.e049167.

Annan, K., 2005. Kofi Annan's statement to the General Assembly | United Nations Secretary-General. [online] Un.org. Available at: www.un.org/sg/en/content/sg/speeches/2005-03-21/kofi-annans-statement-general-assembly [Accessed 2 August 2022].

Aslam, M. 2014. *Education and Pathways of Change. United Nations Educational, Scientific and Cultural Organizations*. University of London. pp.1–69.

Becker, G., 1993. *A Treatise on the Family*. Harvard University Press.

Becker, G.S., 1993. *Human Capital: A Theoretical and Empirical Analysis with Special Reference to Education*, 3rd Edition. The University of Chicago Press.

Bhowmik, J., Biswas, R. and Hossain, S., 2021. Child marriage and adolescent motherhood: a nationwide vulnerability for women in Bangladesh. *International Journal of Environmental Research and Public Health*, 18(8), p.4030.

Copestake, J. G., Brody, A., Greeley, M., and Wright-Revolledo, K. 2005. *Money with a Mission: Microfinance and Poverty Reduction* (Vol. 1). ITDG Publishing / Practical Action Publishing.

Dahl, R., 1957. The concept of power. *Behavioral Science*, 2(3), pp.201–215. https://doi.org/10.1002/bs.3830020303

Das, P., 2019. *Econometrics in Theory and Practice*. Springer.

Debnath, R., Acharjee, L. and Wang., 2019. Empowering women through microcredit in bangladesh: an empirical study. *International Journal of Financial Studies*, 7(3), pp.37.

Doepke, M. and Tertilt, M., 2009. Women's liberation: what's in it for men?. *Quarterly Journal of Economics*, 124(4), pp.1541–1591.

Duflo, E., 2012. Women empowerment and economic development. *Journal of Economic Literature*, 50(4), pp.1051–1079.

Dutta, P., 2025. Bridging the gender gap in education success: Primary education policy in Bangladesh. *Public Administration and Policy*, 28(2), pp.1–13. https://doi.org/10.1108/PAP-10-2023-0138

Education Scenario in Bangladesh: Gender Perspective. 2017. Bangladesh Bureau of Statistics, UCEP Bangladesh Diakonia, Bangladesh. [online] Available at: https://bbs.portal.gov.bd/sites/default/files/files/bbs.portal.gov.bd/page/4c7eb0f0_e780_4686_b546_b4fa0a8889a5/BDcountry%20project_final%20draft_010317.pdf.

Enders, W., 1995. *Applied Econometric Time Series*. USA: John Wiley & Son.

Forgues-Puccio, G. and Lauw, E., 2021. Gender inequality, corruption, and economic development. *Review of Development Economics*, 25(4), pp.2133–2156.

Goldin, C. and Katz, L., 2002. The power of the pill: oral contraceptives and women's career and marriage decisions. *Journal of Political Economy*, 110(4), pp.730–770.

Government of the Peoples Republic of Bangladesh., 1997. Laws of Bangladesh. [online] Available at: http://bdlaws.minlaw.gov.bd/volume-31.html [Accessed 3 August 2022].

Granger, C. W. J., 1969. Investigating causal relations by econometric models and cross- spectral methods. *Econometrica*, 37, pp.424–438.

Haddad, L., Hoddinot, J. and Alderman, H., 1997. *Intrahousehold Resource Allocation in Developing Countries*. Baltimore, MD: Johns Hopkins University Press.

Hashemi, S., Schuler, S. and Riley, A., 1996. Rural credit programs and women's empowerment in Bangladesh. *World Development*, 24(4), pp.635–653. https://doi.org/10.1016/0305-750X(95)00159-A

Haque, M.M., Islam, T., Tareque, M.I. and Mostofa, M., 2011. Women empowerment or autonomy: a comparative view in bangladesh context. *Bangladesh e-Journal of Sociology*, 8(2), pp.17–30.

Haque, R., Alam, K., Rahman, S. M., Keramat, S. A., & Al-Hanawi, M. K. (2021). Women's empowerment and fertility decision-making in 53 low and middle resource countries: A pooled analysis of demographic and health surveys. *BMJ Open*, 11(6), e045952. https://doi.org/10.1136/bmjopen-2020-045952

Heath, R., 2014. Women's access to labor market opportunities, control of household resources, and domestic violence: evidence from Bangladesh. *World Development*, 57, pp.32–46.

Hornset, N. and de Soysa, I., 2022. Does empowering women in politics boost human development? An empirical analysis, 1960–2018. *Journal of Human Development and Capabilities*, 23(2), pp.291–318.

Hossain, N., 2020. The SDGs and the empowerment of Bangladeshi women. In S. Chaturvedi, H. Janus, S. Klingebiel, X. Li, A. de Mello e Souza, & E. Sidiropoulos (Eds.), *The Palgrave Handbook of Development Cooperation for Achieving the 2030 Agenda* (pp. 453–474). Palgrave Macmillan.

Hossain, M. and Tisdell, C., 2005. Closing the gender gap in Bangladesh: inequality in education, employment and earnings. *International Journal of Social Economics*, 32(5), pp.439–453.

International Labour Organization. (2015). The changing nature of jobs. International Labour Organization. Available at: https://www.ilo.org/wcmsp5/groups/public/---dgreports/---dcomm/---publ/documents/publication/wcms_368626.pdf

Jayachandran, S. and Lleras-Muney, A., 2009. Life expectancy and human capital investments: evidence from maternal mortality declines. *Quarterly Journal of Economics*, 124(1), pp.349–397.

Jensen, R., 2012. Do labour market opportunities affect young women's work and family decisions? Experimental evidence from India. *The Quarterly Journal of Economics*, 127, 753–792.

Kabeer, N., 2001. Conflicts over credit: re-evaluating the empowerment potential of loans to women in rural Bangladesh. *World Development*, 29(1), pp.63–84.

Kabeer, N. 2005. Gender equality and women's empowerment: a critical analysis of the third millennium development goal 1. *Gender & Development*, 13, 13–24.

Kabeer, N., 2017. Economic pathways to women's empowerment and active citizenship: what does the evidence from Bangladesh tell us? *The Journal of Development Studies*, 53(5), pp.649–663.

Kabeer, N. and Huq, L., 2014. The power of relationships: money, love and solidarity in a landless women's organization in rural Bangladesh. In A. Cornwall and J. Edwards (Eds.), *Feminisms, Empowerment and Development*, pp.250–276. London: Zed Press.

Kabeer, N., Mahmud, S. and Tasneem, S., 2017. The contested relationship between paid work and women's empowerment: empirical analysis from Bangladesh. *The European Journal of Development Research*, 30(2), pp.235–251.

Kabeer, N., Milward, K. and Sudarshan, R., 2013. Organising women workers in the informal economy. *Gender & Development*, 21(2), pp.249–263. DOI: 10.1080/13552074.2013.802145

Malhotra, A., Schuler, S. R., & Boender, C. 2005. Measuring women's empowerment as a variable in international development. In D. Narayan (Ed.), *Measuring Empowerment: Cross-disciplinary Perspectives* pp.71–88. Washington, DC: The World Bank.

Mamun, K. A. A. and Nath, H. K., 2005. Export-led growth in Bangladesh: a time series analysis. *Applied Economics Letters*, 12(6), pp.361–364.

Mikkola, A. and Miles, C., 2007. Development and gender equality: Consequences, causes, challenges and cures (Discussion Papers / Helsinki Center of Economic Research, No. 159). Helsinki Center of Economic Research. Available at: http://ethesis.helsinki.fi/julkaisut/eri/hecer/disc/159.

Naila, K., Lopita, H. and Simeen. 2014. Diverging stories of "missing women" in South Asia: is son preference weakening in Bangladesh?. *Feminist Economics*, 20(4), pp.138–163. DOI: 10.1080/13545701.2013.857423

Nkoro, E. and Uko, A., 2016. Autoregressive Distributed Lag (ARDL) cointegration technique: application and interpretation. [online] Econpapers.repec.org. Available at: https://econpapers.repec.org/article/sptstecon/v_3a5_3ay_3a2016_3ai_3a4_3af_3a5_5f4_5f3.htm [Accessed 23 July 2016].

Qian, N., 2008. Missing women and the price of tea in China: the effect of sex-specific earnings on sex imbalance. *Quarterly Journal of Economics*, 123(3), pp.1251–1285.

Osmani, L., 2007. A breakthrough in women's bargaining power: the impact of microcredit. *Journal of International Development*, 19(5), pp.695–716.

Oster, E. and Millett, M. 2010. Do call centers promote school enrollment? Evidence from India (NBER Working Paper No. 15922). National Bureau of Economic Research. https://doi.org/10.3386/w15922.

Rahman, L. and Rao, V., 2004. The determinants of gender equity in India: examining Dyson and Moore's thesis with new data. *Population and Development Review*, 30, pp.239–268.

Rose, E., 1999. Consumption smoothing and excess female mortality in Rural India. *Review of Economics and Statistics*, 81(1), pp.41–49.

Salway, S., Jesmin, S. and Rahman, S., 2005. Women's employment in urban bangladesh: a challenge to gender identity?. *Development and Change*, 36(2), pp.317–349.

Schuler, S., Hashemi, S., Riley, A. and Akhter, S., 1996. Credit programs, patriarchy and men's violence against women in rural Bangladesh. *Social Science & Medicine*, 43(12), pp.1729–1742.

Schultz, T.P., 1989. *Benefits of Educating Women*, Background Papers Series, Education and Employment Division, Population and Human Resources Department. Washington, DC: World Bank.

Schultz, T.P., 1990. Testing neoclassical model of labour supply and fertility. *Journal of Human Resources*, 25(4), pp.599–634.

Sen, A.K., 1981. *Poverty and Famine: An Essay on Entitlement and Deprivation*, Oxford: Clarendon Press.

Sen, A.K., 1987. *Gender and Cooperative Conflicts*. Wider Working Papers. Helsinki: World Institution for Development Economic Research, pp.5–60.

Sen, A.K., 1999a. Enhancing women's choices in responding to domestic violence in Calcutta: a comparison of employment and education. *The European Journal of Development Research*, 11(2).

Sen, A.K., 1999b. *Development as Freedom*. Oxford: Oxford University Press.

Sen, A.K., 2001. Gender equity and the population problem. *International Journal of Health Services*, 31(3), pp.469–474.

Shahnaj, S. and Leonhäuser, I., 2005. Empowerment of rural women in Bangladesh: a household level analysis. Abstracts of German Tropentag. Berlin: Humbold-University, pp.5–7.

Shorrocks, A. and Sen, A., 1981. Poverty and famines: an essay on entitlement and deprivation. *Economica*, 50(197), pp.104.

Shrestha, M. and Bhatta, G., 2018. Selecting appropriate methodological framework for time series data analysis. *The Journal of Finance and Data Science*, 4(2), pp.71–89.

Solotaroff, J.L., Aphichoke K., Lonnberg, T., Ali, S., Rohini P. and Ferdous J., 2019. Voices to choices: Bangladesh's journey in women's economic empowerment. Washington, DC: World Bank Group.

Sundström, A., Paxton, P., Wang, Y. and Lindberg, S., 2017. Women's political empowerment: a new global index. *World Development*, 94, pp.321–335.

Tisdell, C. and Hossain, M., 2003. Fertility and female work force participation in bangladesh: causality and cointegration. *Social Economics Policy And Development* (Working Paper).

United Nations. 2022. Gender equality is key to achieving the MDGs: Women and girls are central to development. United Nations Chronicle. Available at: https://

www.un.org/en/chronicle/article/gender-equality-key-achieving-mdgs-women-and-girls-are-central-development

Worldbank.org., 2018. Bangladesh | Data. [online] Available at: https://data.worldbank.org/country/bangladesh.

World Bank, 2021. Labor force participation rate, female (% of female population ages 15+) – Bangladesh. World Bank Open Data. https://data.worldbank.org/indicator/SL.TLF.CACT.FE.ZS?locations=BD

World Bank, 2023. Achieving gender equality is necessary to further reduce poverty and inequality. Washington, DC: World Bank Group. Available at: https://documents1.worldbank.org/curated/en/099042423212573129/pdf/P1767380c5625809408f0c08ea45dc70daf.pdf

World Economic Forum, 2022. Global gender gap report 2022. World Economic Forum. https://www.weforum.org/publications/global-gender-gap-report-2022/

World Economic Forum, 2025. Global gender gap report 2025. World Economic Forum. Available at: https://www.weforum.org/publications/gender-gap-report-2025

Wasse, H., Kutner, N., Zhang, R. and Huang, Y., 2007. Association of initial hemodialysis vascular access with patient-reported health status and quality of life. *Clinical Journal of the American Society of Nephrology*, 2(4), pp.708–714.

Yaya, S., Odusina, E. and Bishwajit, G., 2019. Prevalence of child marriage and its impact on fertility outcomes in 34 sub-Saharan African countries. *BMC International Health and Human Rights*, 19(1), 33.

8 Barriers to Women's Participation in the Bangladeshi Tech Industry

Conceptual Model toward Bridging the Gap

Samiha Mamun and Md. Rahat Khan

8.1 INTRODUCTION

Women entrepreneurship is the ability of women to create, start, operate, and manage firms while facing high levels of social, cultural, and market constraints. This type of entrepreneurial activity contributes to gender equity and improves the economic status of women since it provides full independence and enables women to contribute to the common pool of resources (Kabeer, 2012). The scope of the tech industry captures the industry involved in creating technologies, such as software firms, information technology companies, manufacturing of hardware products, and telecommunication firms. This type of industry is critical for inventing and rejuvenating existing economic structures that enhance contemporary society (Aziz et al., 2023). From a social perspective, digital technology has increased society's likeliness and affected lives. In addition, it has fashioned a new economy that most firms have adopted to sustain themselves in the current world (Samara et al., 2023). According to the World Bank's collection of development indicators, almost half of Bangladesh's population is female, and this data was compiled from officially recognized sources (Chowdhury et al., 2023). Yet the participation of women in the technological field is much less than male participation. This imbalance is occurring due to various barriers encountered by the women of Bangladesh in contemporary times (Saifuddin et al., 2019).

However, it is undeniable that the participation of women needs to be enhanced to foster innovation and enhance technology, which is widespread in the country. The women population also drives economic growth. Thus, the common barriers faced by them include male-dominant cultures (Chandan, 2017), practices like gender inequality in organizations (Natsvlishvili, 2017), and insufficient mentoring opportunities (Whitlock et al., 2023). Western contexts such as the USA, UK, Canada, and Australia have faced similar issues, and plenty of research has been conducted in these countries (Kaplan & Malach-Pines, 2010; Hampton et al., 2011; Kovaleva

DOI: 10.1201/9781003544838-8

et al., 2023), but in South Asia, a country like Bangladesh needs some attention, where the influence of gender roles on sociocultural norms are very common (Pio & Syed, 2013).

According to Industrial Development Leasing Company of Bangladesh (IDLC) in 2023, the female workforce was only 12% in Bangladesh's information technology (IT) sector, occupying low-level positions. Although many significant projects were initiated to increase female participation through various technical education, training, and programs through projects like "She Power" and "Leverage on Information and Communication Technologies (LICT)," gaps remained in the tech sector. Among the fresh female graduates from the Information and Communication Technology (ICT)-based departments or courses, barely 13% moved their careers to the ICT industries. Although the enrollment of female students in the undergraduate and graduate programs was 25%, it was also comparatively less. More attention should be given to this field to focus on the solutions to this imbalance (Murshid, 2023). Economic factors add to these difficulties since women globally experience low financial capability, denying them access to education and training. Becoming a professional becomes even more unachievable due to many factors that include workplace discrimination; women receive lower wages compared to men. These challenges are system-driven and are rooted in the society with biases and cultures that compel women to avoid the sectors seen as dominated by male personnel, such as science, technology, engineering, and mathematics (STEM), which are the core of the technology sector (Kuschel et al., 2020).

Furthermore, while the issue of women being largely underrepresented in the sectors linked to high technologies is a hot topic for scientific research, the reasons that cause such a situation are questioned, including cultural and social factors, gender-role expectations, as well as limits to mobility and safety concerns (Rao, 2005). Even then, these barriers are highly evident in South Asian countries such as Bangladesh due to the high level of patriarchal culture that hinders women's career progression (Raghuvanshi et al., 2017). These barriers relate respectively to Bangladesh's gendered organizational culture and individual concerns for safety and moving around for women. Bangladesh is pushing toward becoming a middle-income economy, and it is high time to prioritize the fight against unemployment by supporting gender-diversified entrepreneurship (Ferdousi & Mahmud, 2019). Women have a responsibility in the process of economic development; thus, it is high time to consider it important to eliminate barriers to women's employment and to ensure favorable conditions in the context of the technology industry.

This study aims to identify and showcase the main factors hindering women from engaging in the technology industry in Bangladesh by understanding the sociocultural, economic, educational, and workplace barriers. In addition, the study aims to develop a conceptual model that presents realistic and contextual solutions to help close these gaps. This research provides an optimal viewpoint on the reasons impeding women's professional advancement in the high-tech sector by examining meso-organizational practices and micro-individual phenomena. It is hoped that this model can be used as a framework to improve participation for women in Bangladesh by informing policymakers, educators, and industry personnel to work toward making study and workplace environments more friendly for women to pursue technology careers. The following sections will cover related theories and literature, methodology, analysis and findings, discussion, and conclusion on how the situation affects gender equality in the country's technology sector.

8.2 RELATED THEORIES AND LITERATURE

8.2.1 Women Empowerment

Empowering women involves enabling them to recognize gender-based disparities in power and equipping them with a stronger platform to advocate against inequality in domestic, professional, and community settings. It entails women asserting autonomy: establishing personal objectives, acquiring competencies, addressing challenges, and fostering self-sufficiency. The viewpoint of feminist empowerment within the International Labor Organization (ILO) framework is to advocate for the equitable status of women in the workforce and to advance this objective at all levels (ESCWA, 2024). In the current age of liberalization and globalization, women have emerged as a formidable workforce, considerably contributing to global economic progress. Consequently, regulators in any economy cannot exclude the significant contributions of women. Equally, advancing the female workforce is prioritized to establish their role as active contributors to economic and societal success. The progress and robustness of a country can be accurately assessed by the collective contributions of both males and females. Women participate in the workforce for different reasons, including glass ceiling challenges in business, work adaptability, societal standing, self-awareness and identity, and the desire to affirm their existence and role within the family, community, culture, and nation (Agarwal & Lenka, 2018).

In Bangladesh, especially, a patriarchal culture plays an important role in determining women's position and their chances. It also explains female subordination in decision-making in matters such as courtship, including employment decisions, before marriage and also after marriage. A deeply patriarchal society significantly influences women's roles and opportunities. Female roles remain subordinate to the male-dominated decision-making in pre- and postmarriage contexts, which prejudices their career liberty (Chowdhury et al., 2023). In addition, patriarchal control prevails within public areas, causing experiences like restricted movement and harassment from some men. Sociocultural practices vary in interpretation and can accommodate inequality between males and females (Gupta, 2020). While Article 28 of the Constitution of the People's Republic of Bangladesh (1972) grants equal rights for the sexes, asserting these rights is still feeble because of the social norms supporting male dominance (Rayhan & Khan, 2020).

8.2.2 Bangladeshi Women and Tech Industry

Bangladesh is swiftly transitioning into a digital landscape, mirroring global trends. As digital and technology breakthroughs proliferate, so do women's possibilities in technology-related professions. The emergence of a digital economy, propelled by governmental initiatives, has increased women's participation in labor and engagement in more demanding roles. According to a survey by the Oxford Internet Institute (OII), Bangladesh is presently the second-largest provider in the online labor market, rendering it one of the nation's most significant socioeconomic instruments. Approximately 500,000 engaged freelancers are consistently engaged from 650,000 authorized freelancers in Bangladesh, a considerable proportion of whom are women. They are producing an estimated yearly revenue of 100 million dollars, as reported by the ICT Department of Bangladesh (Kabir, 2023). An organization like Bangladesh Women in Technology (BWIT) supports women in the technology field by providing training opportunities for intercommunication. Such a step meets the modern approach to the fact that the creation and support of communities is necessary to address the challenges faced by women in STEM. BWIT is a committed forum that helps advance women in the technology industry via training and interconnection. This initiative seems reasonable given that modern thinking includes a determination that building communities and strong networks is important for dismantling the barriers women in technology face (BWIT, 2024). However, the lack of support from family, friends, and even schools, along with the critically low presence of female-inspiring figures – only 22% of women can name at least one famous female in tech – discourages young women from pursuing a career and leads to feelings of loneliness and a perceived inability to belong (Breßler et al., 2023).

8.2.3 Digital Public Infrastructure (DPI) and Bangladesh

The digital public infrastructure (DPI) theory asserts that digital services must be available to all to increase participation in a diverse population, such as women (Sandhu et al., 2023). According to the DPI theory, there is a need to develop more responsive websites to increase the representation of women. This approach acknowledges the understanding that enhancing women's technological competency will go a long way in eliminating the hurdles experienced by women in technology-related areas (Gupta & Nair, 2023; Purandare, 2024). The "She Power" is one of the government plans to provide women with training support and essential tools in the ICT arena. The project was launched in 2018 and has trained thousands of women from different districts, tackling gender inequalities and promoting women's opportunities in the technologies. She Power is a governmental project that aims to involve women in the ICT field and provide them with training and materials. This modern structure is specifically designed to propel the opposite gender and level the playing field at technology-based organizations, improving the odds of women (ICE Business Times, 2021).

8.3 METHODOLOGY

The current research uses a paradigm that is qualitative in nature and uses both survey strategies and case study approaches to identify the factors that hinder employed women in the Bangladeshi tech sector. On the same grounds, qualitative research methodology can be useful in describing and explaining why and how women are gaining employment in this sector and the experiences that they encounter toward the achievement of this employment, as it embraces the ability to provide a detailed picture of the complex phenomenon in the social stratum (Khan & Sarkar, 2024).

8.3.1 Case Selection

The study's research design is presented as a single-case investigation, in which the Bangladeshi tech industry is chosen as the specific context in which such constraints are most effectively identified. The selection includes different ventures, including technology companies and organizations from Bangladesh and, more importantly, those driven or operated by women professionals and business leaders. All the data selected for this case were chosen deliberately, with prior research interviews collected from secondary sources of the interviewees, to capture the nature and extent of gendered challenges that women in this sector experience (Khan & Arif, 2023).

8.3.2 Research Design

The study focuses on secondary research methods: data from published interviews of 16 women professionals working in the Bangladeshi tech industry and 16 published documents from Bangladeshi national dailies, journal articles, magazines, and blogs. The primary method is to include female leaders, entrepreneurs, and advocates in published interviews to discuss their experiences, issues, and views on women's exclusion in the tech industry (Khan, 2019, 2021). The qualitative approach enables the exploration of individual stories, and systematic problems that structures with inherent biases can easily mask through more structured methods (Khan et al., 2022). Moreover, some published materials were utilized in documents to prove the robustness of the analysis concerning women's representation in the tech field (Khan, 2020). The above approaches are intended to seek subtleties that define the challenges that hinder female participation in the Bangladeshi technological sector and strategies that provide solutions to such hurdles.

8.3.3 Data Collection and Analyzing Procedures

Data collection uses qualitative and quantitative means, combining interviews and document reviews to validate the observations (Khan & Sharma, 2020). The research was conducted with judgmentally collected published interviews of 16 female participants, leader entrepreneurs, and activists and 16 published documents related to the Bangladeshi tech industry to get insights and views on the barriers to women's participation in the Bangladeshi tech industry (Khan, 2020; Malhotra, 2020). The interviews are also intended to generate both observational and descriptive data to enhance the odds of understanding women's challenges in the tech industry (Khan et al., 2022). This systematic approach enables greater investigation of the antecedents to female engagement while also capturing entrenched structures of the technology industry within Bangladesh. The study developed a process model for facilitating women's participation in the Bangladeshi tech industry by addressing existing hurdles. The research employed NVivo (Version 11) and MS Word (Version 2019) software to develop the model through cognitive assessment (Table 8.1).

TABLE 8.1
Summary of Data Collection

Informant Number	Informant's Affiliation/ Occupation	Experience (Years)	Interview Duration (Minutes)
I-1	BWIT Member	10	45
I-2	Freelancer	5	30
I-3	Robotics Department	7	40
I-4	Engineer	8	35
I-5	Vocational Training Advocate	6	50
I-6	Financial Institution	12	42
I-7	Entrepreneur	4	38
I-8	Energy Sector Professional	10	36
I-9	University Educator	9	45
I-10	Gender Equality Advocate	11	33
I-11	University Student & Job Holder	4	30
I-12	E-commerce Entrepreneur	5	40
I-13	ICT Forum Member	6	35
I-14	Tech Start-up	7	50
I-15	Community Organizer	4	32
I-16	Govt. Official	8	48

Documents and Observations

Documents Type	Quantity	Sources
Blogs	10	Google Search
Newspaper Articles	6	*The Business Post, Prothom Alo, The Business Standard*, and *The Financial Express*

Source: Authors.

8.4 ANALYSIS AND FINDINGS

Drawing on published interviews and documents, the qualitative study identified key findings about why women are discouraged from entering Bangladesh's tech industry. Respondents included technology leaders, educators, and business professionals. Thematic analysis revealed both the systemic challenges women face and actionable strategies to increase their participation in the sector.

8.4.1 Word Frequency Analysis and Generation of Themes

All recorded data are consolidated into a Microsoft Word document and imported into NVivo (Version 11) software for qualitative data analysis to study the frequency of words and create a word cloud for thematic coding. The study selected terms with a minimum size of six letters for word frequency assessment, focusing on the most frequently used 25 words, with the results summarized in Table 8.2. In word cloud evaluation, the research selected terms with a minimum size of four letters, and Figure 8.1 illustrates the general findings of the word cloud.

This particular analysis guided the research in developing themes to compile the data coding and structure in the next section of this research.

8.5 DISCUSSION AND THE PROCESS MODEL

Based on the above analysis, the following conceptual model can be proposed to learn the barriers to women's participation in the Bangladeshi tech industry and bridge the gap and the way to success. The model depicts a systematic issue where women face complex barriers to equal participation. The process model delineates four critical steps: family and financial support; society, gender biases in the workplace, and technophobia as barriers to the process; and the overcoming of the issues that may fuel the success of women's participation in the Bangladeshi tech industry.

TABLE 8.2
Word Frequency Summary

SN.	Frequency	Word	Similar Words
1	222	Women	Female, Woman, Woman's, Girl, Gender
2	97	Business	Project
3	35	Create	Idea
5	58	People	Students
6	74	Journey	Working, Start
7	44	Bangladesh	Country
8	44	Entrepreneurs	Entrepreneur
9	39	Technology	Tech
10	23	Participation	Engagement
11	21	University	College, School
12	44	Support	Help, Money
13	30	Digital	Electronic, Computing, Use, STEM
14	18	Believe	Faith, Understanding
15	17	Family	Parents
16	17	Opportunities	Scope
17	27	Social	Society, Life
18	14	Energy	Ability
19	14	Challenges	Difficult
20	13	Equality	Equal
21	12	Talking	Talk
22	12	International	Global
23	12	Training	Train
24	23	Progress	Success
25	40	Education	Learn

Source: Authors.

FIGURE 8.1 Word cloud analysis from interview outcomes.

Source: Authors.

8.5.1 Step 1: Education (Requires Family Support)

Familial support is crucial in motivating women to pursue education and enroll in educational institutions for IT studies. This step establishes the foundation for acquiring the requisite skills and knowledge to enter the technology sector. The result underscores the need for women to have parental encouragement and transcend traditional norms dictating appropriate occupations for their gender. Two crucial interview related to this identified theme are shared below to learn the importance of education for women in the tech industry.

> Women's engagement in the ICT sector must be enhanced, necessitating the integration of relevant skills from a young age, ideally beginning in basic education.
>
> In the present educational framework, students in Bangladesh are required to select between science, business education, or humanities disciplines upon finishing eighth grade. Parents significantly influence this decision-making process. Studies indicate that female students from rural regions frequently encounter financial obstacles in pursuing education in the science field. Parents may emphasize spending in their sons' studies and their daughters' marriages over their girls' education in the sciences. This trend is prevalent nationwide, posing a substantial obstacle to attaining gender equality in STEM education despite certain exceptions.

Education becomes the initial and the most important enabling factor of the model. The interviews showed that barriers start in institutions where social prejudices and parental involvement make their first steps. Such representation hinders young girls from aspiring for a career in technology, resulting in a so-called cycle of less representation of girls in technology fields (Huang, 2017; Aithal & Maiya, 2023). The above statements mean that with the help of educational facilities that introduce the principles of inclusiveness and ensure the growth of interest in STEM from a very young age, there will always be an increased number of women pursuing technology courses (Dökme et al., 2022). Possible positive impacts arising from enhanced education include increases in a skilled workforce capable of addressing issues in the technological sector.

8.5.2 Step 2: Innovation and Knowledge (Requires Financial Support)

There is a need for financial capital concerning education and innovation to enable women. In any of these sectors, they can take up training, skill development, and other programs to improve their technical skills and fill the gap between their studies and the job market. Innovation and knowledge advances the educational dimension by insisting on knowledge about the concepts and practices that fund ideas and accompanying activities. The following interview is noted to learn the role of innovation and expertise in women's careers in the tech industry.

> Innovative technologies are crucial in advancing human rights, particularly women's rights.
>
> I obtained my degree in agricultural science to tackle the challenges confronting the farming community and was resolute in applying my academic expertise and enthusiasm to effect change.
>
> I was without employment. "At that moment, a friend proposed that I initiate a business based on a straightforward concept. It appeared as though conducting business was an exceedingly simple endeavor. Thus, I commenced my adventure into entrepreneurship within the IT sector.
>
> The concept of transitioning to an online platform is very challenging. I did not attend school, so I am unfamiliar with online education. Alternatively, I may have engaged in online commerce.
>
> Perhaps educated people can apply their knowledge and participate in work online. Working in the tech industry without education and knowledge is unfeasible for individuals like us."

Creativity is not contingent upon financial resources (Heunks, 1998). Conversely, innovation necessitates financial resources. Financial assistance's beneficial impact in facilitating funding for innovation is enhanced when innovators have exceptional knowledge of innovation (Clò et al., 2022). The following interview from the analysis shows how well the financial support assists.

> Furthermore, if you are a woman, your circumstances will be very intense. When a female entrepreneur applies for a bank loan, bankers frequently perceive that the funds are intended not for the lady herself but instead for her father or spouse. Occasionally, a spouse may secure a loan under his wife's name; nonetheless, genuine entrepreneurs should not bear the consequences of another's transgressions. Financial institutions are tasked with recognizing genuine entrepreneurs and providing them with services. It is never acceptable to close the door in someone's face due to their gender. Such practice is also unethical and a barrier to women's exploration speed.

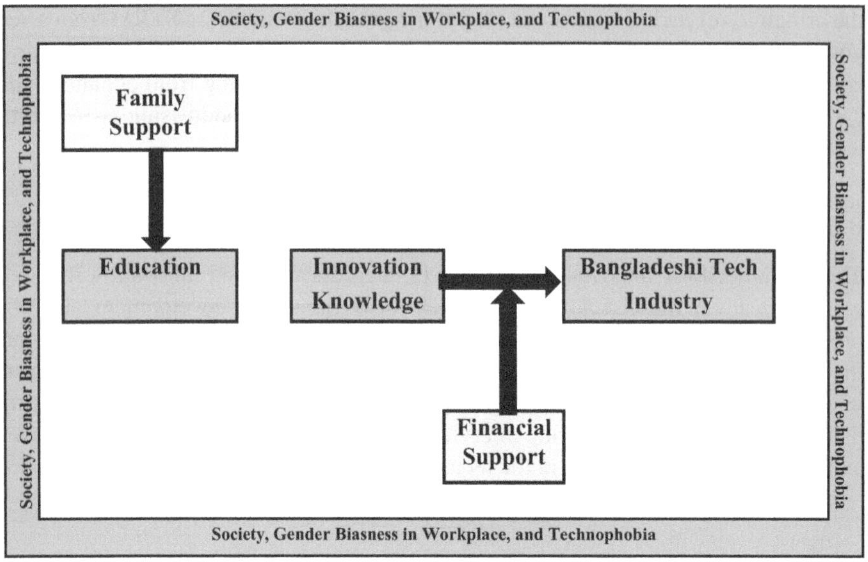

FIGURE 8.2 Process model on overcoming barriers to women's participation in the tech industry.

Source: Developed by the authors.

Education and financial support work hand-in-hand; a good idea might need capital to be actualized. Overcoming this problem by devoting specific funds to support women could make them turn knowledge into good tech solutions, creating a great environment for further development (Haider, 2023).

8.5.3 Step 3: Society, Gender Biasness in the Workplace, and Technophobia (Barriers to the Process)

Based on the analysis, the observed major barriers to the process of women's flourishing in the Bangladeshi tech industry are society, gender bias in the workplace, and technophobia. Therefore, it is identified that women encounter major barriers during the process. After receiving an education and financial support for their endeavors, women are still limited by cultural and social practices, employment discrimination, and fear of or against technology (technophobia). These barriers have to be brought down to increase women's engagement in the tech industry. Thus, it is evident that when the Bangladeshi education and innovation channel are launched, and the barriers to women's access to the sector are addressed, women can efficiently enter the Bangladeshi tech Industry. In this case, success involves equality, participating in leadership roles, and adding value and creativity.

Societal influences and success underline the role of ongoing prejudices faced by women in society, female marginalization at the workplace, and the fear of sexophobia (Kabir et al., 2018). Interviewees from the informants described how people challenge women's career capacity.

> Women must exhibit a lack of shame. A culture that denies a woman her rightful opportunities is a disgrace to that community. A woman must emerge and dismantle the silence, barriers, and societal conventions; otherwise, transformation would remain unattainable.
>
> Challenges will perpetually persist. Society will invariably oppose change unless we confront the established norms. Risk will perpetually exist, and circumstances will never be ideal; we must advance and pursue our desires.

They can limit women's advancement despite educational attainment and idea creation without these societal barriers. If women are still being discriminated against in areas concerning employment, the same or similar job prospects reduce dramatically (Cooke et al., 2019). In addition, women's technophobia in the tech industry is another barrier that can be defeated to increase participation and success in the tech industry.

> People frequently underestimate my programming abilities since it is commonly believed that women gravitate toward marketing, client management, or, at most, teaching.
>
> The term "online" is unfamiliar to us. Significant competition, fraudulent activities, and a lack of integrity in business practices constrain online selling.

The research findings indicate that women's participation in technology necessitates collaborative efforts at the micro, meso, and macro levels via education, policy, and cultural transformation activities. Consequently, strategies targeting each phase of the model may be implemented, and by efficient dissemination of initiatives among relevant stakeholders, a foundation for the effective empowerment of women and sustainable advancement of the technology sector in Bangladesh can be established.

8.5.4 Step 4: Success in the Bangladeshi Tech Industry

Reducing barriers indicates that when barriers are reduced across education, innovation knowledge, and societal perceptions, the female representation in the tech industry could significantly improve. A few statements from the informants and study analysis are pointed to as evidence.

> I reject the notion that business is inherently a masculine profession. It is how our civilization is projected. Regrettably, the business system is formed around this specific projection in Bangladesh. However, the landscape is rapidly evolving, with numerous women joining the realm of entrepreneurship. Women are capable of conducting business, and I exemplify this fact.
>
> I commenced my journey with programming and then transitioned to analysis and proposition for project design, which is more demanding than societal perceptions suggest a woman can undertake.

While overcoming obstacles, positive impacts include increased female representation in technology careers and increased innovation by diverse inputs, thus resulting in a diverse workforce that increases economic returns. On the other hand, the failure to attend to the barriers will cause the achievements of half the population to be subdued, with the concomitant effect of holding back the economy's growth (Clancy et al., 2018).

8.6 CONCLUSION, IMPLICATIONS, AND FUTURE DIRECTIONS

The research has revealed key aspects of the complex situation that affects women in this industry. The study shows that there has been an increased understanding of the current and potential role of women in computer science, although there is a persistent gap between demand and supply. The conceptual model formulated through this study helps identify these barriers, which are not isolated but are interrelated, and which impact women's participation in the tech sector. The study has several limitations: firstly, informants were quite defensive of their experiences because of their privacy rights and because the topic was sensitive. Secondly, several participants, especially those from the business community, proceeded with overwhelming caution to share detailed information; their responses were more corporate and polished than reflecting their struggles. Finally, due to the exploratory nature of this study, some of the descriptive findings of assumed barriers need further research.

Some implications of this study are that readers will appreciate the practical significance of this research, going beyond theory and applying it to the different stakeholders promoting women in technology. In particular, the results highlight the importance of adopting policies that can reduce the financial constraints that have been found to impede women participants or entrepreneurs from achieving their full potential. These are matters of fostering and providing special loans and funding for women's technology firms. Also, the formulation of norms requiring software developers and technological firms to employ women can create policies that would actively protect women in those positions.

Schools can mold future generations of women technologists in any country. According to the study, there is a need to encourage STEM courses and the use of practical assignments as part of girl's curriculum from a young age. Through projects that promote technology, schools can encourage girls against the stereotype that will sway them from STEM fields. In addition, stronger links between academic institutions and private sector organizations can create internship and employment opportunities for female students. Lastly, community organizations and advocacy groups must keep spreading awareness regarding gender in technology. Awareness can be achieved through awareness creation programs directed at families that aim to alter the general perception of women's capability to take up tech careers.

The study has helped establish awareness of women's involvement in the technological field without necessarily exploring much of the causality relation of the different factors. For instance, demographic factors such as age, education level, and experience – key factors determining women's entrepreneurship – were not well explored. More research could consider these factors as moderators to better understand their effect on women's experience in the technology field. Finally, in line with the literature reviewed in this research, concrete barriers to women's representation in the Bangladeshi tech industry have been identified while encouraging more extensive research to study the potential ramifications of the topic. While the conceptual model helps conceptualize how these barriers are connected, an additional quantitative investigation must be conducted to determine its general impact on women's economic enfranchisement and the tech industry's development. Nevertheless, this study provides the initial base for the subsequent research and policy efforts to increase women's representation in the technology industry.

The study concludes that collective endeavors by different institutions, that is, governmental departments, school administrators, and employers, will play a crucial role in women's nurturing. The interviews thus established that measures taken to enhance women's engagement must go beyond the basic problem-solving approach and embrace the perspective of building a new culture. Bangladesh can add value by focusing its attention on solving problems at the ergonomic level and creating effective environments for the country's workforce, thus providing a more stable and long-term solution to the problem. Therefore, this study adds relevant knowledge to the existing literature on gender in technology by reviewing female workers' challenges in Bangladesh.

REFERENCES

Agarwal, S., & Lenka, U. (2018). Why research is needed in women entrepreneurship in India: A viewpoint. *International Journal of Social Economics*, *45*(7), 1042–1057.

Aithal, P. S., & Maiya, A. K. (2023). Innovations in higher education industry–Shaping the future. *International Journal of Case Studies in Business, IT, and Education (IJCSBE)*, *7*(4), 283–311.

Aziz, G., Sarwar, S., Nawaz, K., Waheed, R., & Khan, M. S. (2023). Influence of tech-industry, natural resources, renewable energy and urbanization towards environment footprints: A fresh evidence of Saudi Arabia. *Resources Policy*, *83*(2023), 103553.

Breßler, J., Hegemann, L., & Huesig, S. (2023, July). Analyzing the Role of "Women in Technology" for Digitally Transformed Organizations. In *2023 Portland International Conference on Management of Engineering and Technology (PICMET)* (pp. 1–12). IEEE.

BWIT. (2024). *About us.* Retrieved from www.bwit.org.bd/

Chandan, H. C. (2017). *Motivations and challenges of female entrepreneurship in developed and developing economies.* In *Discrimination and Diversity: Concepts, Methodologies, Tools, and Applications* (Vol. 3, pp. 1613–1640). Edited by Information Resources Management Association. IGI Global

Chowdhury, S., Rahman, M. M., & Haque, M. A. (2023). Role of women's empowerment in determining fertility and reproductive health in Bangladesh: A systematic literature review. *AJOG Global Reports*, *3*(3), 100239.

Clancy, S. M., Stroo, M., Schoenfisch, A., Dabrera, T., & Østbye, T. (2018). Barriers to engagement in a workplace weight management program: A qualitative study. *American Journal of Health Promotion*, *32*(3), 763–770.

Clò, S., Frigerio, M., & Vandone, D. (2022). Financial support to innovation: The role of European development financial institutions. *Research Policy*, *51*(10), 104566.

Cooke, D., Fernandes, A. P., & Ferreira, P. (2019). Product market competition and gender discrimination. *Journal of Economic Behavior & Organization*, *157*, 496–522.

Dökme, İ., Açıksöz, A., & Koyunlu Ünlü, Z. (2022). Investigation of STEM fields motivation among female students in science education colleges. *International Journal of STEM Education*, *9*(1), 1–14.

ESCWA. (2024). *Women's empowerment.* Retrieved from www.unescwa.org/sd-glossary/women%E2%80%99s-empowerment

Ferdousi, F., & Mahmud, P. (2019). Role of social business in women entrepreneurship development in Bangladesh: Perspectives from Nobin Udyokta projects of Grameen Telecom Trust. *Journal of Global Entrepreneurship Research*, *9*(1), 58.

Gupta, A., & Nair, A. (2023). *Unpacking Digital Public Infrastructure: Navigating Conceptual Ambiguities. T20 Policy Brief.* Observer Research Foundation.

Gupta, N. (2020). Rationalizing gender inequality at scientific research organizations: A reproduction of the Indian socio-cultural context. *Equality, Diversity and Inclusion: An International Journal*, *39*(6), 689–706.

Haider, M. (2023). Behind women's emancipation and oppression: Contested expertise in global microcredit governance. *Australian Feminist Law Journal*, *49*(2), 239–265.

Hampton, A., McGowan, P., & Cooper, S. (2011). Developing quality in female high-technology entrepreneurs' networks. *International Journal of Entrepreneurial Behavior & Research*, *17*(6), 588–606.

Heunks, F. J. (1998). Innovation, creativity and success. *Small Business Economics*, *10*(1998), 263–272.

Huang, S. P. (2017). Effects of innovative education on innovation capability and organizational performance in high-tech industry. *Eurasia Journal of Mathematics, Science and Technology Education*, *14*(3), 777–784.

ICE Business Times. (2021). *She Power Project-Putting the Power in Her Hands*. Retrieved from https://ibtbd.net/she-power-project-putting-the-power-in-her-hands/

Kabeer, N. (2012). Women's Economic Empowerment and Inclusive Growth: Labour Markets and Enterprise. *School of Oriental and African Studies, UK SIG.*

Kabir, S. (2023). *Bangladeshi women in tech and digital education: Challenges and strides.* Retrieved from www.thedailystar.net/supplements/international-womens-day-2023/news/bangladeshi-women-tech-and-digital-education-challenges-and-strides-3266051

Kabir, S. M. S., Aziz, M. A., & Shathi, A. S. J. (2018). Women empowerment and governance in Bangladesh. *ANTYAJAA: Indian Journal of Women and Social Change*, *3*(1), 24–35.

Kaplan, G., & Malach-Pines, A. (2010). Women Serial High-Technology Entrepreneurs. In A. Malach-Pines & M. F. Özbilgin (Eds.), In *Handbook of Research on High-Technology Entrepreneurs*. Edward Elgar Publishing.

Khan, M. R. (2019). Samson H. Chowdhury: A true entrepreneurial leader for Square Group. *EuroMed Journal of Management*, *3*(2), 80–88.

Khan, M. R. (2020). Ethical business practice and short term earning intensity during the Covid-19 crisis in Bangladesh. *International Journal of Financial, Accounting, and Management*, *2*(2), 79–93.

Khan, M. R. (2021). A critical analysis of Elon Musk's leadership in Tesla motors. *Journal of Global Entrepreneurship Research*, *11*(1), 213–222.

Khan, M. R., & Arif, M. Z. U. (2023). Operational efficiency through disruptive innovation in dairy business: Evidence from an emerging economy. *International Journal of Productivity and Quality Management* (Advance online publication) doi.org/10.1504/IJPQM.2023.10058494

Khan, M. R., & Sarkar, U. (2024). Business leadership in the digital age: IKEA's story in the retail and consumer sector. *Green Technologies and Sustainability*, *2*(2), 100081. doi.org/10.1016/j.grets.2024.100081

Khan, M. R., & Sharma, K. (2020). Purchase preferences and buying influences on religious occasions. *FIIB Business Review*, *9*(3), 216–227.

Khan, M. R., Roy, S. K., & Pervin, M. T. (2022). Retail-based women entrepreneurship entry model through small business orientation (SBO). *Journal of Women's Entrepreneurship and Education*, *1-2*(2022), 117–136.

Kovaleva, Y., Hyrynsalmi, S., Saltan, A., Happonen, A., & Kasurinen, J. (2023). Becoming an entrepreneur: A study of factors with women from the tech sector. *Information and Software Technology*, *155*, 107110.

Kuschel, K., Ettl, K., Díaz-García, C., & Alsos, G. A. (2020). Stemming the gender gap in STEM entrepreneurship–Insights into women's entrepreneurship in science, technology, engineering and mathematics. *International Entrepreneurship and Management Journal*, *16*(1), 1–15.

Malhotra, N. K. (2020). *Marketing Research: An Applied Prientation*. Pearson.

Murshid, K. (2023). *She Power: A tribute to my mother Begum Nurjahan Murshid*. Retrieved from www.thedailystar.net/op-ed/she-power-tribute-my-mother-begum-nurjahan-murshid-87847

Natsvlishvili, I. (2017). Gender Inequality and Women's Entrepreneurship-Challenges and Opportunities (Case of Georgia). In M. H. Bilgin, H. Danis, E. Demir, & U. Can (Eds.), *Country Experiences in Economic Development, Management and Entrepreneurship: Proceedings of the 17th Eurasia Business and Economics Society Conference* (pp. 491–505). Springer International Publishing.

Pio, E., & Syed, J. (2013). Our bodies, our minds, our men: Working South Asian women. *Gender in Management: An International Journal*, *28*(3), 140–150.

Purandare, U. (2024). Digital infrastructure: Public good or private commodity?—Rethinking the right to internet access in the context of urban digitisation. *Science, Technology and Society*, *29*(4), 555, 09717218241281824. doi.org/10.1177/09717218241281824

Raghuvanshi, J., Agrawal, R., & Ghosh, P. K. (2017). Analysis of barriers to women entrepreneurship: The DEMATEL approach. *The Journal of Entrepreneurship*, *26*(2), 220–238.

Rao, M. (Ed.). (2005). *Knowledge Management Tools and Techniques: Practitioners and Experts Evaluate KM Solutions*. Routledge.

Rayhan, A., & Khan, T. I. (2020). The Constitution of People's Republic of Bangladesh-Solemn Expression of People's Will; A legal basis of social changes. *Indian Journal of Law & Justice*, *11*(1), 234–263.

Saifuddin, S., Dyke, L., & Hossain, M. S. (2019). Walls all around: Barriers women professionals face in high-tech careers in Bangladesh. *Equality, Diversity and Inclusion: An International Journal*, *38*(7), 705–726.

Samara, E., Andronikidis, A., Komninos, N., Bakouros, Y., & Katsoras, E. (2023). The role of digital technologies for regional development: A system dynamics analysis. *Journal of the Knowledge Economy*, *14*(3), 2215–2237.

Sandhu, K., Dayanandan, A., & Kuntluru, S. (2023). India's CBDC for digital public infrastructure. *Economics Letters*, *231*(2023), 111302.

Whitlock, D. W., Hampton, A. J., & Campbell, K. (2023). Mentoring women entrepreneurs: Perceived needs and benefits. *Journal of Business & Entrepreneurship*, *33*(1), 1–21.

9 Women Empowerment in Tech Startups of Bangladesh

Government Initiatives and Internet Penetration

Apoorva Thakur

9.1 INTRODUCTION

The word "entrepreneurship" often denotes the creation of new enterprises. In developing nations, entrepreneurship is the paramount strategy for economic growth and sustainability. Entrepreneurs are pivotal in every economy because they use their talents, skills, and efforts to exploit underutilized resources. They serve as the catalysts for a nation and function as the foundational elements of economic development. Entrepreneurs own the largest share of business in any industrialized nation and significantly contribute to gross domestic product (GDP). Women have become significant contributors in the realm of entrepreneurship (Saif, 2021).

The economic empowerment of women is an essential component of development rhetoric. The institutionalization of sustainable development processes that exclude women from mainstream development programs is untenable and will hinder sustainable economic progress. Nonetheless, sustained progress in commerce and development would remain an elusive objective without the active and significant involvement of women in the nation. According to the latest sex ratio data, women make up more than half of Bangladesh's population. They constitute an important part of the country's pool of potential employees. There is an absolute need for women to participate actively in the mainstream business due to the demographic structure. It is impossible to have a thriving and sustainable economy unless women, who make up half of the population, are actively and meaningfully involved in it (Debnath et al., 2020).

In order for a country to thrive and prosper, and for democracy to take root, there must be a sustainable economy. Incorporating women into the formal sector is crucial if we want to build a society free of poverty. There has been a concerted effort by the government of Bangladesh, non-governmental organizations (NGOs), the corporate world, and other relevant parties to increase the number of women working in official economic sectors, especially in manufacturing and commerce. United States Convention to Eliminate All Forms of Discrimination Against Women (CEDAW) was

DOI: 10.1201/9781003544838-9

accepted by Bangladesh and other states. In all spheres of public life, the Constitution of Bangladesh guarantees equal rights for men and women (Articles 28(1) and 28(3)). According to Article 10 of the Constitution, the state must ensure that women participate actively and significantly in all spheres of public life. As a result of constitutional mandates and international pressure, the government of Bangladesh has established institutional measures to raise women's economic and social status. The government has taken several important steps so far, such as establishing a separate ministry for women's affairs, formulating the National Policy for Advancement of Women—2008, and responding to the Beijing Platform for Action (PFA) with the National Action Plan (Mazumdar & Alharahsheh, 2020).

The United Nations Development Program (UNDP) Human Development Report (2004) defines a woman entrepreneur as someone who has started or inherited a business, either alone or with others, and is prepared to take on social, administrative, financial, and risky responsibilities, in addition to managing day-to-day operations. A rising nation such as Bangladesh cannot contemplate economic advancement while maintaining half of its populace in unemployment. Currently, the Bangladeshi government has instituted many measures to foster and propagate entrepreneurship among women in both urban and rural regions. Although the prevalence of women-owned enterprises is minimal compared to those owned by males, it is encouraging as it indicates that women are no longer constrained by the notion that only men serve as the primary earners of the family. Women may achieve success by establishing their enterprises, although their responsibilities include several domestic tasks, including reproductive duties such as housework and childcare (Sultana, 2012).

9.2 STATUS OF WOMEN ENTREPRENEURSHIP IN BANGLADESH

Women's entrepreneurship undoubtedly enhances national income and familial prosperity. Contemporary women are more inclined to engage in pursuits formerly seen as male domains, demonstrating their equal contribution to economic progress. Women's entrepreneurship must be effectively shaped with requisite entrepreneurial qualities and abilities to adapt to evolving trends, address global market issues, and maintain competitiveness while pursuing excellence in the entrepreneurial domain.

McKinsey Global Institute analysis in 2015 (a) hints at a playing out of a situation that eradicates the gender gap can add as much as 28 trillion USD to the global GDP by 2025 and (b) a boost in female employment rates can absorb 240 million employees. Trade might play a very influential role in this process. However, several impediments today exist today to women across the world depriving them from attaining economic independence through commercial activities (Rahman, 2016).

Still, many challenges affect women entrepreneurship in Bangladesh. Stereotyping of the genders especially the male and female still persists and is a worrying feature of Bangladeshi culture. Men in this world fail to put much value on women's achievements at home, in society, and in the nation. Female business owners repeatedly work to negotiate their worth through overcoming social barriers. The factors relating to the economy and society of Bangladesh cause women's entrepreneurship controversy. They face many challenges like limited funding, insufficient experience, illiteracy, cultural prejudice, familial discouragement, ignorance, lack of access to the markets, lack of knowledge and how-to, technical assistance, and religious superstitions (Shoma, 2019).

However, the nation has witnessed the generation of a new kind of woman entrepreneur willing to face the odds of operating in a competitive, patriarchal, and demanding economic environment. Women entrepreneurs have a better status of living and enhanced their recognition among their families and other communities. They were now literate of their social rights and has been actively practicing the chance given to them by different government and non governmental agencies. The observation of women in economical activities within Bangladeshi culture has experienced social change. The level of awareness of men and partners in making the woman productive appears to be better and men are willing to support them. This change has been made possible by government regulations, identifying the role of financial institutions, both public and private, and other support services (Akter et al., 2019).

The Bangladesh Bank provides prospective banks with recommendations for prioritizing women in the allocation of small- and medium-sized enterprise (SME) loans. Simultaneously, several non-profit microfinance institutions (NGO-MFIs) offer loans to women engaged in micro, small, and medium enterprises. Research by the Bangladesh Bureau of Statistics (BBS) indicates that women's access to land, credit, and other assets has increased by 0.80, 0.30, and 0.50%, respectively (BBS, 2009). The mobilization of resources, including agricultural entrepreneurs (cattle, fisheries, and crops), tailors, small company proprietors, and artisans, among others, motivated rural women to pursue entrepreneurship (Bangladesh Regional Connectivity Project, Ministry of Commerce, 2018).

A BCG-Mass Challenge investigation indicates that investments in firms established or co-established by women averaged USD 935,000, which is less than half of the USD 2.1 million average spent in companies started by male entrepreneurs. This disparity persists despite women-founded startups demonstrating superior performance, yielding 10% greater cumulative revenue over five years, with revenues of USD 730,000 for women-founded startups versus USD 662,000 for men-founded startups. This discovery is a significant consideration in investing in women-led enterprises in Bangladesh (Rana, 2022).

Female entrepreneurship, particularly in the technology sector, remains an undeveloped domain in Bangladesh. The Mastercard Index of Women Entrepreneurs for 2020 ranked Bangladesh 58th out of 58 economies. The nation fell one position from last year's 57th, while its total score increased to 36.4 out of 100 from 35.4. The index's accompanying study asserts that entrenched socio-cultural, economic, and financial barriers, such as insufficient employment opportunities, governmental assistance, and access to funding and capital, continue to impede women in these economies (Haque et al., 2023).

The Mastercard Index includes the women's advancement outcomes component (Component A). The component serves as a proxy for women's advancement and the extent of their economic and professional marginalization as business leaders, professionals, entrepreneurs, and members of the labour market. In Bangladesh, women's advancement as trained professionals and their involvement in the labour force often lag behind their worldwide counterparts. Although female professionals constitute over 30% of all occupations, the female-to-male ratio in the workforce remains dishearteningly low at roughly 3:10. Bangladesh has one of the lowest proportions of female company owners, with 4.5% of overall business owners. Furthermore, Bangladesh achieved the lowest ranking in Component B: Knowledge Assets and Financial Access. This component evaluates the advancement of women, the extent of their marginalization as financial clients, and their access to university education enrollment. It assesses women's propensity to borrow or save money for entrepreneurial endeavours, as well as the extent of support provided to SMEs regarding government assistance and foundational infrastructure (Afrin et al., 2024).

Component C ranks the entrepreneurial circumstances as factors that either help or hinder women's capacity to advance and succeed as company owners; Bangladesh fared rather well (in comparison to other components) in this area, being placed at the 55th position (The Daily Star, 2022). Additionally, it assesses the role of socio-cultural factors in promoting or discouraging female entrepreneurship. The survey asserts that women in this area have significantly fewer opportunities to advance professionally as skilled workers and assume leadership roles in businesses, compared to their counterparts in other locations. A lack of established physical infrastructure and government initiatives to support their economic endeavours also makes their lives more difficult. Since the epidemic began, Facebook has established more than 70% of female-led enterprises in Bangladesh, while Instagram has seen an increase in female-owned businesses by over 65%, according to Meta. For a long time, women's social standing in Bangladesh has been the focus of heated controversy. There have been significant gains for women in terms of employment and educational achievement, but there are still many obstacles that prevent them from fully participating in decision-making. Newer industries like IT also face these difficulties. On the other hand, women may find new options for work and a chance to break down old gender stereotypes in the information technology industry. Over the last decade, Bangladesh's information technology industry has grown at an astounding rate, creating new possibilities for women in areas like education, training, and employment. However, the industry is not without its challenges. Persistent barriers to advancement for women in information technology include discrimination, inadequate funding, and an absence of female role models and mentors (Khanum et al., 2020).

9.3 ACCESS TO TECHNOLOGY VITAL FOR WOMEN ENTREPRENEURSHIP DEVELOPMENT

Entrepreneurship advancement and its sustainability entail technology. As a source of initial help in overcoming barriers to communication, it reduces the time needed to accumulate data, recognized as a vital resource for an entrepreneur. Besides, it minimizes costs and increases the competitiveness of an enterprise.

In 2016, internet penetration in Bangladesh was 14%, an increase from 10% in 2015. In 2016, internet penetration in China reached 50%. Technology significantly impacts corporate operations. Regardless of an enterprise's size, technology offers both real and intangible advantages that assist women entrepreneurs in generating revenue, achieving outcomes, and satisfying client demand. Moreover, technology infrastructure influences the culture, efficiency, and relationships inside a company or farm. The Sustainable Development Goals (SDGs) aim to transform the 21st century by tackling critical issues such as poverty, inequality, and violence against women and girls. Empowering women is a prerequisite for this. SDG 5 seeks to achieve gender equality and empower all women and girls, especially by enhancing the use of enabling technology, particularly information and communication technology (ICT) to promote women's empowerment. An application is now accessible to assess the amount of time a woman dedicates to work and her corresponding earnings, addressing digital literacy gaps. Technology may enhance the affordability of a superior quality of life, while an entrepreneur can improve production efficiency by using demand-orientated technology. Bangladesh still lacks strong policy support for women entrepreneurs using digital technology, despite evidence that ICT can reduce business costs and enhance competitiveness; recent initiatives by the Business Initiative Leading Development (BUILD) aim to strengthen policy advocacy in this sector (The Financial Express, 2017; BUILD, 2024). Initially, we understood that women entrepreneurs primarily operate in the micro and cottage sectors, concentrating on traditional, small-scale economic handicrafts and fashion-oriented textile goods. Currently, women are increasingly participating in outsourcing by acquiring freelance training in software development, electrical and electronics engineering, effects, lighting engineering, and several other fields. Bangladesh, with a population of 170 million and around 147 million mobile customers, is a significant market in Asia. Bangladesh is a nation of opportunities for businesses (United Nations ESCAP, 2022).

As Bangladesh strives to accelerate economic development, enabling women to launch new ventures becomes essential for strengthening both the economy and society. Orser, Riding, & Li (2019) emphasizes that access to technology not only offers advantages to women entrepreneurs but also helps them overcome gender bias, geographic barriers, and limited resources, thereby enabling greater participation in entrepreneurial activities. In order to support women entrepreneurs in Bangladesh to realize their potential, gain efficiencies, reach new clients, and encourage innovation, technology needs to be folded into business pursuits. However, some hurdles such as digital knowledge, cost, and technology accessement remain a big challenge for women business owners. In a bid to reaching out and encouraging more women to participate in business and, therefore, boost the economic status of the country, the above concerns must be dealt within the framework of special policies and programmes (Thorsteinsdóttir et al., 2021).

9.3.1 The Role of Technology in Women's Entrepreneurship

Greater use of technology makes available a number of tools and instruments that can be useful for small- and medium-sized enterprises today. Technology holds promise for the women entrepreneurs in Bangladesh facing several challenges, such as limited access to funds and restrictions of socio-cultural nature. This also opens new markets for companies whose operations it assists in digital marketing campaigns, while increasing resource utilization by automating and optimizing corporate processes. One of the most significant advantages of technology is it creates markets irrespective of how exclusive they may be. For example, women business owners may use e-commerce to break the geographical limitations and get a wider market access. When you think of the fact that a good number of women business entrepreneurs are located in villages within Bangladesh and not close to urban centers, you will realize the significance of this. Stores such as Daraz and Evaly are now offering thousands of women the chance to sell products online cutting out the costs associated with physically having a store. Research by LightCastle Partners revealed that about 68% of the internet merchants in Bangladesh are women, most of whom are perhaps micro business owners. Thanks to such platforms, they have been able to take their businesses to the next level and reach more people, hence they earn more money and have more access to finances. In addition, social networking sites, specifically Instagram and Facebook, have become mandatory for Bangladeshi women entrepreneurs. Through these channels businesses are able to promote their products at a cheap cost, directly engage with the consumer, and nurture the customer base (Sultana & Akter, 2021).

9.3.2 Digital Financial Services and Access to Credit

Women entrepreneurship lacks adequate access to financial services in Bangladesh. Commercial banks traditionally discriminate against women because of strict security and enforcement of collaterals and gender biases in loan-granting processes. These challenges are being addressed by technology with the takeoff of digital finance and mobile money services.

One such includes the bKash mobile banking facility which has improved magnitude of access to funds by the vulnerable groups such as women. We hope that advancement in mobile wallets and digital payment systems have reduced challenges that businesswomen had when accessing banking services. According to the findings of Bangladesh Bank as of 2023, the number of registered accounts under mobile banking services is 104 million and women are contributing a significant sector of mobile banking service users. The more flexible digital financial services are benefiting self-employed women by allowing them to control money on their own and reducing their vulnerability to intermediaries (Moral et al., 2024).

In addition, Bangladesh-based technology-friendly microfinance organizations such as Bangladesh Rural Advancement Committee (BRAC) and Grameen Bank are incorporating digital technologies into their lending operations. These banks offer business loan portfolios to women, many of whom own micro or small enterprises. In the case of microfinancing, which concerns the provision of small amounts to poor people to sustain their small businesses, the opportunities of digitization contribute to better distribution and control of the lending and reduce administrative burdens in making financial products available (Sarker, 2020).

9.3.3 Digital Literacy and Training Programs

While availability of technology may help catalyse the growth of women – empowered businesses in Bangladesh, literacy still remains an issue. There are still many women, especially in rural areas, who have insufficient skills among modern technology for their businesses. According to the BBS of Bangladesh, the current literacy rate of women is 71.9% while digital literacy is considerably less. The digital divide affects women's uptake on e-commerce platforms and digital financial services which puts a check on their capacity to create and sell in the market (Emon & Nipa, 2024).

In turn, various initiatives and governmental activities concerning female entrepreneurship have been developed with the goal to raise their digital competence. Bangladesh Government's ICT Division has started projects similar to "She Power Project", which aims to provide education on computers, business management and internet marketing to women. UNDP-supported Access to Information (a2i) Program has proved central in eliminating the digital literacy gap through its numerous training courses for women entrepreneurship (Rakib, 2023).

Such programs are supported by non-governmental organizations (NGOs), commercial organizations, and public sector undertakings (PSUs), which work in collaboration with these institutional structures. Grameen Intel Social Company Ltd. offers training solutions for women entrepreneurs to support the efficiency of the companies with techniques of digital social skills like stock control and digital advertising.

9.3.4 Government Policies and Infrastructure

It demonstrates that government policy is equally important as availability for enabling women entrepreneurs to get access to technology. The government of Bangladesh has proposed a plan to get the women linked with digital platforms in its latest 2021 Digital Bangladesh Vision. The ICT Policy made in 2018 stresses the requirements for the increasing of females' participation in the ICT sector and the assuring of technology networks to everybody including businesswomen. The specific sector reveals the government having spent considerable resources in developing the digital base across the country. The National Broadband Policy is aimed to offer affordable broadband connectivity on the most remote areas in Bangladesh. The government is improving the broadband delivery to help women exporters penetrate the global market and to take advantage of the emerging e-commerce business. Furthermore, the applied financial inclusion measures of Bangladesh Bank are primarily targeting women entrepreneurs. Women entrepreneurship development program provides a concessional finance facility to women-led SMEs sector. This goals of the initiative are to encourage the banks and other financial institutions to extend the credit facilities that favour women owned businesses as well as to provide the tools in the areas of accounting and technology (Ripa et al., 2023).

9.4 GENDER INCLUSIVENESS IN THE COUNTRY'S TECHNOLOGY SECTOR

In order for IT businesses to make better and safer products that include everyone, not just one group, gender equality and diversity are crucial. Deloitte, a professional service network located in London, projects that by 2022, the percentage of women working in top technology businesses globally will have dropped to 25%. Bangladesh has a much more dismal situation when it comes to women's engagement in the IT sector. The Business Standard reports that in 2020, only 12% of IT jobs were held by women, with the majority of those women holding entry-level or middle-level positions. Nonetheless, it is no longer a secret that diversity and gender equality are of essential importance (Afrin et al., 2024).

In 2020, women made up 34.55% of Bangladesh's workforce, according to the World Bank. From 2000 to 2020, the rate of women participating in the labour market has been significantly increasing, according to the International Labour Organization (ILO). The gender pay gap has been narrowing and women's empowerment has been on the rise in Bangladesh, a country in South Asia. The rate has increased twofold in a short period of time. The World Economic Forum (WEF) has announced that, for the seventh year running beginning in 2014, Bangladesh has ensured more gender equality than any other South Asian country. Dhaka Tribune reports that a 3.1% boost to GDP is possible if Bangladesh can achieve a 10% increase in female employment. The budget for FY2021–22 has prioritised policies and initiatives that support women, which has led to a surge in women joining the workforce (Moin, 2024).

In order for society and the economy to progress, women must play an essential role. With hard work and perseverance, many women have reached the highest levels of leadership in well-known companies, while others have discovered their calling in business. Thanks to globalization, more women are able to take advantage of the opportunities presented to them and join the workforce. Women now have the self-assurance and skills to show their worth on a larger stage. They have assumed leadership roles and are now representing the nation abroad.

Technology enhances the degree of efficiency, which in turn drives economic development in this age of digitalization. Similar to other nations, Bangladesh has embraced technology and is now heavily dependent on it. Since 2010, the information and technology-enabled services industry in Bangladesh has been projected to increase at a rate of 40% per year, according to LightCastle Partners. The telecommunications, banking, pharmaceutical, and textile industries in Bangladesh have recently seen extensive automation efforts. The introduction of novel goods to consumers is a direct result of technological advancements in the banking industry. Digital transformation of fast-moving consumer goods (FMCG) has shown encouraging development despite the pandemic, thanks to technology. Domestic demand for the software and Information Technology Enabled Services (ITES) sectors is anticipated to rise further for growth and efficiency.

Tech Observer reports that by the conclusion of FY2021–22, one million young people have found work in the information technology industry in Bangladesh, a country with a youthful population that exceeds 50% of the total. For those with scientific and technological backgrounds, the government's "Sustainable Development Goals, Vision 2041" and "Delta Plan 2100" have made it easier to find work (Lata, 2024).

9.4.1 Initiatives Taken to Promote Women's Empowerment in Tech

In this age of rapid technological development, it is common knowledge that a nation's economic prospects and human capital may be enhanced by fostering an environment that encourages women to pursue careers in science and technology. Consequently, efforts have been made by both the governmental and commercial sectors to encourage women to take part in Bangladesh's technology economy. What follows is a discussion of a few of the efforts (Anjum et al., 2024).

LICT Project—Through a World Bank-funded training initiative called Leveraging ICT for Growth, Employment and Governance (LICT), the local government has made efforts to encourage females to pursue careers in information and communication technology (ICT) with the goal of achieving a minimum participation rate of 30%. Last year, on 30 June, 10,450 women were trained as part of a project that aimed to produce more than 10,000 women who could work as competent IT human resources.

She Power—For the purpose of advancing women in information and communication technology, the government of Bangladesh has launched the "She Power" initiative. In 2019, 10,500 women from 21 districts were taught by the ICT Division as part of this project.

Bangladesh Women in Technology (BWIT)—Bangladesh Women in Technology (BWIT) has been initiated as a dedicated platform aiming at promoting women in the technology industry through training and a diverse range of networking opportunities with members including ICT entrepreneurs and professionals in the ICT, computer science, EEE, robotics and mechatronics sectors.

The government's "Digital Bangladesh" strategy is only one of several initiatives aimed at empowering women in Bangladesh. In addition to them, organizations like the "Bangladesh Computer Council" (BCC) and the "Bangladesh Korean Institute of Information and Communication Technology" (BKIICT) have worked to empower women in Bangladesh. The Bangladesh Bureau of Statistics (BBS) reports that in 2020, 36.3% of Bangladeshi women were working. Nonetheless, only 15.8% of women have official employment, a far smaller proportion than the male-dominated workforce. Domestic work, street selling, and agriculture make up the bulk of Bangladeshi women's employment in the informal sector. Individuals in the formal sector hold the majority of technological occupations (Sobhan & Hassan, 2024).

Even the RMG sector, where women in Bangladesh have achieved enormous success, has failed them in several ways. Women made up 59.4% of RMG employees but held less than 2% of management or ownership positions, according to a 2019 Survey of Manufacturing Industries (SMI) survey that surveyed over 8,000 RMG firms. Statistics on women's access to technology provide evidence of women's participation in Bangladesh's tech workforce (Solotaroff et al., 2019). In 2022, BBS conducted a Survey on ICT Use and Access by Individuals and Households 2022, revealing that 72.3% of males and 51.4% of females possess mobile phones, an SDG indicator. Similarly, 45.3% of males and 32.7% of females have used the internet in the last three months, also an SDG indicator. Women constitute less than one-third of the global workforce in STEM occupations, according to a World Bank analysis. In the STEM fields, women make up 28% of the workforce, whereas they account for 15.9% of the engineering and architectural workforce (Mimi et al., 2022). Among European Union (EU) workers in the information and communication technology field, women constitute only 17%. Examining women's roles in the traditionally male-dominated IT sector is pertinent as we enter a new decade. There is a gender disparity in the IT business because many organizations have been slow to implement inclusive hiring procedures. Furthermore, women face higher rates of gender-based discrimination and lower earnings compared to males (Thorsteinsdóttir et al., 2021).

Microaggressions are commonplace for women in the workplace, according to a WomenTech Network poll with over 600 respondents. Approximately 64% of those who took the survey had experienced being ignored or talked over while in a meeting, 19% have been stereotyped in different contexts and 11% have been asked to "supply the food" while in a meeting. According to the poll, many people consider gendered language and taking meeting minutes to be "the usual" forms of microaggressions.

For Smart Bangladesh to thrive in the future, the government must prioritize policies and programs that empower women economically. This should include setting a goal of having 30% female leadership positions in both the public and private sectors by 2030. In light of this, the Smart Bangladesh ICT strategy contains specific goals to eliminate the gender gap, such as establishing an internal "Gender gap task force" to promote women's participation in smart device access programs and to protect them when they shop online. The strategy aims to establish a digital job platform that will provide equal opportunities for women in industries where their employment rates have historically been lower. Education Minister Dr. Dipu Moni approved the new curriculum earlier this month, which begins teaching kids to code in Grade III and continues through high school. Education Minister Dr. Dipu Moni has scheduled its implementation in all elementary and secondary schools by 2025. In Bangladesh, women face several obstacles while trying to enter the technology industry, one of which is the widespread lack of educational opportunities. Many programs exist to help girls and women get the technical skills and STEM degrees they need, but these efforts need to spread to all the places where they are most necessary.

One of the obstacles to women's participation in the work field in Bangladesh is our mentality. A more progressive outlook is also necessary for Smart Bangladesh if we want to see more women join the traditionally male-dominated field of technology. Software engineers, computer science majors, STEM fields, and skilled occupations are all open to women, but they face discrimination and harassment in the workplace due to outdated cultural attitudes and biased hiring practices. This can make it challenging for them to advance in their careers (Islam, 2016).

A large number of strong, accomplished women from Bangladesh are breaking into the IT business and are already building a name for themselves. We urgently need a platform that assists women in technology in finding mentors, applying for fellowships and scholarships, building a network of peers, and accessing professional development tools. It is critical to implement educational programs that inspire more young women to pursue careers in technology. A more equitable Bangladesh, where all citizens may benefit from technological advancements and have a voice in national affairs, is the goal of Smart Bangladesh. In addition, the IT sector has to make an effort to empower women in senior positions via various programs. In the next decades, Bangladesh will grow to prominence as Smart Bangladesh thanks to the participation of women in technology and tech-enabled enterprises. We think that when more women work in tech, it will lead to a more inclusive workplace. By working together, we are empowered to impact the lives of our nation's future citizens.

In 2008, the government of Bangladesh unveiled *Vision 2021*, a strategy to achieve middle-income status by the year 2021—the country's 50th anniversary of independence. Empowering women, especially by supporting their entrepreneurial endeavours, was a vital component of this initiative's various aims. The unrealized potential of female entrepreneurs was the motivating force behind Vision 2021, which sought to create conditions in which women could break down social and economic barriers and become the main drivers of economic progress. Initiatives associated with this endeavour are closely connected to the UN's Sustainable Development Goal 5: Gender Equality and Women's Empowerment (Andriamahery & Qamruzzaman, 2022).

This information may be linked to experienced businesswomen in Bangladesh, particularly those managing small and medium enterprises. The absence of banking credit, market intelligence, networking opportunities, and other finance sources has hindered their ability to expand their enterprises or into new areas. For example, it is important to acknowledge that, in pursuit of Vision 2021, which aims to create a more inclusive economy, women in business are also encountering similar problems.

Facilitating finance access for female entrepreneurs was a primary goal of the Vision 2021 initiative. In Bangladesh, like in almost other regions worldwide, women have considerable hurdles when seeking a loan. Lending money without collateral is inherently dangerous; hence, women have historically been barred from traditional banking organizations. The Bangladesh government addressed these difficulties by fostering tight relationships with financial industry organizations under Vision 2021 to regulate women's loans. The central bank of the nation, known as Bangladesh Bank, has implemented numerous initiatives that incentivise commercial banks and microfinance institutions to allocate a certain proportion of their portfolios to women-operated enterprises. However, the introduction of microcredit initiatives for financial aid was especially essential for grassroots women's empowerment. Notwithstanding the modest loans, the women successfully started and managed small enterprises that had considerable importance for the economy, particularly in rural regions. Despite the existence of several microfinance organizations today providing microcredit to individuals in need, pioneers such as Grameen Bank and BRAC have played a significant role in this movement, particularly in extending credit to female entrepreneurs (Debnath et al., 2020).

The creation of capacity-building programs that would enable women company owners to have all that they need when it comes to starting up their businesses, running every aspect of such businesses, and managing growth and expansion was also part of Vision 2021. It was noted that a significantly limited number of Bangladeshi businesswomen appealed for education or experience in business management, particularly in the rural and semi urban areas. The government, with assistance of a few NGOs and private players, came up with trainings which sought to bridge this gap. These programmes underscore issues to do with sound financial management, company management, computer literacy, and market access. These initiatives also align with the Government's vision of building a "Digital Bangladesh" by 2021, with a strong focus on technology. Under Vision 2021, women were focused on for greater inclusion in the formal sector by imparting essential skills to them, such as e-business operations and mobile phone banking. This initiative was expected to expand their entrepreneurial opportunities by helping them reach more clients and grow their businesses (Thorsteinsdóttir et al., 2021).

The results have shown that women business owners have received special gains from digitalization. E-commerce platforms such as Daraz, AjkerDeal, Bikroy, and others who have been assisting the small- and medium-sized enterprises owned by women have become more crucial. In an effort to ensure women can sell assorted items online despite problems such as low earning, lack of transport and other geographical constraints, Vision 2021 supported the creation of online markets. New form of financial services such as mobile financial services—BKash, Nagad—made women more integrated in the economy by providing them easier transactions and payments. Second, the government undertook polices to enhance women's marketing opportunities both in the local and global market under Vision 2021. The International Business Centre (ITC) and other institutions created programs such as "She Trades Commonwealth" that would ensure that women entrepreneurs were able to access foreign markets. By so doing, women have benefited by being in a position to come out of their respective local market segments into the international market segment of trading. Overall these kinds of programs have been helpful to the economy of Bangladesh as they have helped to bring out more women into entrepreneurship and get them endorsed in the international market.

Vision 2021 aims to support businesswomen in the ICT sector, which has been a significant contributor to the Bangladeshi economy. Government training and promotions, as well as starter packs, have been implemented to encourage women to enter this sector. The Women ICT Frontier Initiative (WIFI) has been funded by APCICT (Asian and Pacific Training Centre for Information and Communication Technology for Development), providing women with mentorship and access to networks to excel in fields previously dominated by men. While some barriers have been met, Bangladeshi women in business still face challenges due to cultural beliefs and sentiments that keep them locked out of the economy. However, Vision 2021 measures have created opportunities for women to break free from traditional roles and become significant players in the nation's economic growth by accessing markets, technology, and funds.

9.5 THE BARRIERS TO BECOMING A WOMAN ENTREPRENEUR IN BANGLADESH

Women entrepreneurship is increasing in Bangladesh's modern economy. Women are increasingly establishing enterprises at unprecedented rates. Bangladesh was among the first emerging nations to advocate for women's economic empowerment in 1990 (Begum, 2002). Subsequently, several evaluations and studies have been undertaken to assess women's advancement, obstacles, and opportunities in the entrepreneurial sphere. A significant number were located in certain regions of the nation, such as Dhaka, Khulna, or Sylhet, or focused on particular sectors, including handicrafts or manufacturing. Moreover, several studies only focus on possibilities, while others address challenges; moreover, many of these studies are outdated and fail to represent the current landscape of women's entrepreneurship in Bangladesh (Moral et al., 2024).

Female entrepreneurs in Bangladesh lag behind their male counterparts in economic involvement, empowerment, and leadership. In Bangladesh, enterprises owned and led by women constitute around 7%, a figure that is insignificant relative to the female population of 82.3 million. This amount is incalculable in the sustainable economic development of the nation. The Entrepreneurship Index developed by CEOWORLD Magazine evaluated 100 global economies, representing nearly 95% of the world's GDP, using six key indicators: openness for business, innovation, infrastructure, competitiveness, labour skills, and access to capital. These factors collectively reflect how effectively each economy supports entrepreneurial activity and business growth, highlighting differences in enabling environments across countries (Jones, 2024). A dismal 12.99 out of 100 was Bangladesh's final score. Bangladesh came in dead last out of 58 economies in the 2020 Mastercard Index of Women Entrepreneurs. Consequently, Bangladesh represented the most challenging environment for women to start a company. The societal context and administrative framework in Bangladesh make women's business more arduous. Numerous studies indicate that women significantly contribute to entrepreneurial endeavours and economic expansion by generating employment opportunities and enhancing GDP, so positively impacting poverty alleviation and social inclusion. The quantity of new female entrepreneurs has markedly risen in recent years, positively influencing the workplace, domestic sphere, and community (Emon & Nipa, 2024).

While entrepreneurship is challenging for everybody, it is more arduous for female entrepreneurs in Bangladesh. The decision-making of a female CEO faces challenges both inside and outside just due to her gender.

The emphasis on helping female entrepreneurs often centres on enhancing access to financing (financial capital) and offering training to facilitate skill development (human capital)—two essential domains for augmenting the performance of female-led enterprises. Nevertheless, a crucial element in the success of these enterprises is often neglected: access to networks. Establishing market links is far more challenging for women than for males.

The departure rate for female founders is greater than that of their male counterparts, however, an exit from a company does not always indicate failure. In order to determine what variables impacted the departure choices of 219 male and female Spanish entrepreneurs, Rachida Justo and colleagues conducted an analysis. Rather than leaving their businesses because they failed, female entrepreneurs are more likely to leave willingly and for personal reasons, according to their research. Both male and female entrepreneurs' voluntary departure options are significantly and uniquely influenced by familial factors, such as marital status and motherhood. Justo found that entrepreneurs of both sexes quit for reasons other than financial failure, however the specific reasons differed across the sexes. The probability of women departing for personal reasons increased after marriage. In keeping with traditional gender norms, this classifies women as "caretakers" and males as "family providers". Comparable data for Bangladesh are unavailable; nonetheless, entrepreneurship, formerly seen as a male-dominated sphere, has evolved in the country's setting. Nevertheless, as an increasing number of women enter entrepreneurship, they often encounter hurdles not commonly experienced by their male colleagues. Their principal hurdles include: (1) challenging societal norms; (2) restricting access to capital; (3) acknowledging personal achievements; (4) establishing a support network; (5) reconciling professional and familial obligations; and (6) managing a fear of failure (Sarker et al., 2024).

For a female entrepreneur operating in the same industry as her male equivalent, acquiring consumers may seem challenging. A myth exists suggesting that female entrepreneurs excel primarily in certain sectors, such as boutique retail on Facebook or food-related enterprises. A female founder operating a software company is seen as socially less acceptable.

Access to digital equipment and dependable, cost-effective internet access emerged as a significant obstacle for females in realizing their technological professional aspirations. Philanthropy is crucial in enhancing access, in conjunction with the government's computer lab initiative at educational institutions. Nonetheless, digital literacy remains inadequate owing to a general deficiency in access to devices and the internet.

Security has emerged as a concern for females in the digital realm. Insufficient understanding of safe social media use is rendering women susceptible. There exists a significant deficiency in the comprehension of digital privacy and data security. There is potential for increased participation of women in policy formation and at the policymaking level since their existing engagement is insufficiently substantial (Anjum et al., 2024).

9.6 CONCLUSION

The path towards women empowered within the context of the technology startups of Bangladesh is evolving but challenging. Gradually, yet constantly, substantial improvement has been made in the past few years—emerging from government-led programs to increasing internet usage; however, there are many challenges nevertheless. Through such programmes like "Vision 2021" and "Digital Bangladesh", the barriers in terms of training, financial aid, and growing online facilities have been removed. These have helped in pushing women into tech-driven entrepreneurship; for instance, the Learning and Earning Development Project has trained over 30,000 women. In addition, new-generation women-owned micro and small enterprises and increase of e-businesses like Daraz and Evaly also show how use of technology helps women to explore new arenas in business.

However, gender inequality is still a problem. The fact that women are underrepresented in the formal economy and even the tech industry can be attributed to poor socio-cultural attitudes towards women's employment, lack of access to technology, and financial barriers. Female representation is still low in the IT sector and they still struggle for the following reasons: funding, gender discrimination, and lack of female IT professionals to emulate. The trend is even starker when one looks at startups founded by women, which perform better than male-founded and co-founded startups, but attract investment half as much. This lack of inclusion in financial markets still keeps their potential far from being achieved. In light of this, it is necessary for further efforts to be made in order to close the digital divide and achieve sustainable development of women entrepreneurs vying for a space within the world's technology industry. These efforts include increasing awareness with regard to new technologies, expanding access to technology at an affordable price, and creating equal opportunities in the business environment. The government has an important role to play through its policies such as the "She Power" project, and support from non-governmental organizations and financial institutions is essential to support women in technology leadership; there is need to encourage women to take leadership positions in the technology segment. Although Bangladeshi society is steadily progressing towards a more diverse, digitally connected economy, these factors will only become critical as it develops more. It is not just a matter of recognizing women's rights to work in the tech startups but also of recognizing the significance of women as having potential to add worth to the nation's economy and contribute to innovation.

REFERENCES

Afrin, S., Hoque, M. S., & Akter, B. (2024). Women's leadership development through entrepreneurship in Bangladesh. *Management and Labour Studies*, 49(2), 208–240.

Akter, M., Rahman, M., & Radicic, D. (2019). Women entrepreneurship in international trade: Bridging the gap by bringing feminist theories into entrepreneurship and internationalization theories. *Sustainability*, 11(22), 6230.

Andriamahery, A., & Qamruzzaman, M. (2022). Do access to finance, technical know-how, and financial literacy offer women empowerment through women's entrepreneurial development? *Frontiers in Psychology*, 12, 776844.

Anjum, N., Rahaman, M. S., Rahman, M. M., & Dana, L. P. (2024). From start to success: Women entrepreneurs navigating the entrepreneurial ecosystem in Bangladesh. In L.-P. Dana & M. Chhabra (Eds.), *Women Entrepreneurship Policy: Context, Theory, and Practice* (pp. 31–51). Springer Nature Singapore.

BBS (2009). Bangladesh Bureau of Statistics, Statistical Division, Ministry of Planning, Government of the People's Republic of Bangladesh.

Begum, A. (2004). Protection of Women's Rights in Bangladesh: A Legal Study in an International and Comparative Perspective. *University of Wollongong*.

BUILD. (2024). Current projects: Women Entrepreneurship Development and Policy Advocacy. *Business Initiative Leading Development*. https://buildbd.org/current-projects.

The Daily Star. (2022). Bangladesh: A Tough Place to be a Woman Entrepreneur. https://www.thedailystar.net/business/news/bangladesh-tough-place-bewoman-entrepreneur-2000069

Debnath, G. C., Chowdhury, S., Khan, S., & Chowdhury, T. S. (2020). Achieving sustainable development through entrepreneurship & economic empowerment of women in the technological era. *International Journal of Management*, 11(9), 1385–1398. www.iaeme.com/IJM/issues.asp?JType=IJM&VType=11&IType=9

Emon, M. H., & Nipa, M. N. (2024). Exploring the gender dimension in entrepreneurship development: A systematic literature review in the context of Bangladesh. *Westcliff International Journal of Applied Research*, 8(1), 34–49.

The Financial Express. (2017, November 1). Access to Technology Vital for Women Entrepreneurship Development. The Financial Express. https://thefinancialexpress.com.bd/views/reviews/access-to-technology-vital-for-women-entrepreneurshipdevelopment-1509557620

Haque, A., Haque, W., Ahmed, E., Islam, M. Z., & Islam, M. K. (2023). Empowering women through IT: The role of the IT sector in Bangladesh. *Saudi Journal of Economics and Finance*, 7(10), 459–465.

Islam, K. A. (2016). Impact of Micro-financing on women empowerment in Bangladesh. *Millennium University Journal*, 1(1), 1–18.

Jones, L. (2024). World's Most Entrepreneurial Countries, 2024. https://ceoworld.biz/2024/04/05/worlds-most-entrepreneurial-countries-2024/

Khanum, F., Akter, N., Deep, T. A., Nayeem, M. A. R., & Akter, F. (2020). The role of women entrepreneurship in women empowerment: A case study in the city of Barishal, Bangladesh. *Asian Journal of Advances in Research*, 3(1), 346–357. https://jasianresearch.com/index.php/AJOAIR/article/view/206.

Lata, L. N. (2024). Navigating corporate frontiers assessing barriers and facilitators for women entrepreneurs in Bangladesh. Ministry of Commerce. (2018, December). Bangladesh Regional Connectivity Project-1: First implementation support mission – Aide-mémoire, November 12-19, 2018. Government of the People's Republic of Bangladesh & The World Bank. documents1.worldbank.org/curated/en/793871668439087494/txt/Bangladesh-First-Regional-Connectivity-Project-Implementation-Support-Mission-November-12-19-2018.txt

Mazumdar, A., & Alharahsheh, H. H. (2020). Digital Bangladesh–vision 2021: What is the digital Bangladesh concept. *South Asian Research Journal of Engineering and Technology*, 2(1), 6–9.

Mimi, A., Imran, M. A., Beg, T. H., & Rahman, M. S. (2022). Governmental and institutional initiatives and actions for the attraction and expansion of e-commerce by women in Bangladesh. *American Economic & Social Review*, 9(1), 17–28.

Ministry of Commerce. (2018, December). Bangladesh Regional Connectivity Project December. Ministry of Commerce. (2018, December). Bangladesh Regional Connectivity Project-1: First implementation support mission – Aide-mémoire, November 12-19, 2018. Government of the People's Republic of Bangladesh & The World Bank. documents1.worldbank.org/curated/en/793871668439087494/txt/Bangladesh-First-Regional-Connectivity-Project-Implementation-Support-Mission-November-12-19-2018.txt

Moin, S., & Sarkar, A. (2024). Analysis of the opportunities and challenges of women's entrepreneurship in developing countries: A quantitative study from Pakistan, India, and Bangladesh. (Master's thesis, Halmstad University). Halmstad University.

Moral, I. H., Rahman, M. M., Rahman, M. S., Chowdhury, M. S., & Rahaman, M. S. (2024). Breaking barriers and empowering marginal women entrepreneurs in Bangladesh for sustainable economic growth: A narrative inquiry. *Social Enterprise Journal*, 20(4), 585–610.

Orser, B., Riding, A., & Li, Y. (2019). Technology adoption and gender-inclusive entrepreneurship education and training. *International Journal of Gender and Entrepreneurship*, 11(3), 273–298. https://doi.org/10.1108/IJGE-02-2019-0026

Rahman, S. M. A. (2016). Leveraging ICTs for empowering women entrepreneurs in Bangladesh: A review of recent developments and way forward. *European Journal of Business and Social Sciences*, 5(3), 1–15.

Rakib, M. R. H. K. (2023). Factors discouraging the growth of women entrepreneurship in Bangladesh. *International Journal of Entrepreneurship and Small Business*, 48(3), 215–230.

Rana, Md. Masud. (2022). Diffusion of women entrepreneurship in Bangladesh: Current status, challenges, and opportunities. *NDUB Journal of Business Studies*, 1(1), 83–110.

Ripa, F. A., Mustari, N., Azam, N., Mithila, J. M., & Nayem, Z. (2023). Women entrepreneurs-contributions through the digital marketplace: In the economy of Bangladesh. *American Journal of Youth and Women Empowerment*, 2(1), 1–7.

Saif, S. B. (2021). Technology based entrepreneurship and its prospect in Bangladesh. *International Journal of Trade and Commerce-IIARTC*, 10(1), 190–209.

Sarker, A. (2020). ICT for women's empowerment in rural Bangladesh. Doctoral dissertation, Monash University.

Sarker, M. F. H., Ahmed, S. F., Kawser, U., Hossen, M. S., Obaidullah, M., Khan, S., ... & Sohel, M. S. (2024). Is e-business breaking down barriers for Bangladesh's young female entrepreneurs during the COVID-19 pandemic? A qualitative study. *SN Social Sciences*, 4(6), 107.

Shoma, C. D. (2019). Gender is a human rights issue: The case of women's entrepreneurship development in the Small and Medium Enterprise sector of Bangladesh. *Journal of International Women's Studies*, 20(7), 13–34.

Sobhan, N., & Hassan, A. (2024). The effect of institutional environment on entrepreneurship in emerging economies: Female entrepreneurs in Bangladesh. *Journal of Entrepreneurship in Emerging Economies*, 16(1), 12–32.

Solotaroff, J.L., Kotikula, A., Lonnberg, T., Ali, S., Pande, R.P., Jahan, F. (2019). Voices to Choices: Bangladesh's Journey in Women's Economic Empowerment. United States: *World Bank Publications*.

Sultana, A. (2012). Promoting women's entrepreneurship through SME: Growth and development in the context of Bangladesh. *IOSR Journal of Business and Management*, 4(1), 18–29.

Sultana, F., & Akter, A. (2021). Women e-commerce: Perspective in Bangladesh. *Journal of Management, Economics, and Industrial Organization*, 5(3), 1–13.

Thorsteinsdóttir, H., Bell, J. M., & Bandyopadhyay, N. (2021). Cultivating small and medium-sized firms: Entrepreneurship development, gender, and technology in Bangladesh, Cambodia, Ethiopia and Senegal. United Nations Technology Bank for the Least Developed Countries & Small Globe, Inc.

United Nations Development Program (2004). Human Development Report 2004: Cultural Liberty in Today's Diverse World. *Oxford University Press*.

United Nations Economic and Social Commission for Asia and the Pacific. (2022). Bangladesh startup ecosystem assessment report. UN ESCAP.

10 Women in the Tech Startup Sphere in Bangladesh

Jashim Uddin Ahmed, Zulfiquer Ali Haider, and Amreen Talukdar

10.1 INTRODUCTION

In recent years, the tech startup landscape in Bangladesh has experienced a significant transformation, driven by rapid digitalization, a growing economy, and a burgeoning interest in innovation. Amidst this evolution, women have emerged as pivotal players in the tech startup ecosystem, challenging traditional gender roles and contributing to the industry's growth. Despite facing societal and structural barriers, an increasing number of women entrepreneurs are stepping into the tech startup sphere, bringing fresh perspectives and innovative solutions to the market. The participation of women in the tech startup ecosystem is vital not only for gender equity but also for the overall economic development of Bangladesh. Women-led startups often focus on diverse and inclusive products and services, addressing gaps that have historically been overlooked. These startups are playing a crucial role in driving digital inclusion, particularly in areas like education, healthcare, and financial services which are essential for the country's progress. However, the journey for women in tech startups is fraught with challenges. Cultural norms, limited access to funding, and a lack of mentorship and networking opportunities are significant hurdles that women entrepreneurs in Bangladesh face. Despite these challenges, many women are forging ahead, supported by a growing number of initiatives and programs aimed at fostering female entrepreneurship in the tech sector. As more women break through barriers and establish themselves as leaders in the tech industry, they are not only redefining the entrepreneurial landscape but also paving the way for future generations of women innovators.

10.2 TECH STARTUP ECOSYSTEM

The tech startup ecosystem refers to the interconnected network of entrepreneurs, investors, mentors, incubators, accelerators, and support organizations that collectively supports the growth and success of technology startups (Shareef et al., 2024). This ecosystem plays a crucial role in fostering innovation, driving economic growth, and shaping the future of industries worldwide. At its core, the tech startup ecosystem thrives on innovation and disruption. Entrepreneurs bring forward new ideas and technologies, aiming to solve problems or meet market needs in novel

DOI: 10.1201/9781003544838-10

ways. Incubators and accelerators provide essential support, offering mentorship, funding opportunities, and access to resources that help startups refine their ideas and scale their operations rapidly. Investors are pivotal within this ecosystem, providing the financial backing that enables startups to develop prototypes, conduct market research, and, ultimately, bring their products or services to market. These investors range from angel investors and venture capitalists to corporate venture arms and government funds, each contributing to the financial landscape that sustains startup growth. Mentors and advisors, often experienced entrepreneurs or industry professionals, offer guidance and expertise to startup founders, helping them navigate challenges, make strategic decisions, and avoid common pitfalls. Networking events, conferences, and co-working spaces further enrich the ecosystem by facilitating collaboration, knowledge sharing, and partnerships among startups and stakeholders. Moreover, regulatory policies, access to talent, and infrastructure also influence the tech startup ecosystem's dynamics. Regions with supportive regulatory environments, robust educational institutions producing skilled graduates, and modern digital infrastructure tend to attract and nurture vibrant tech startup communities.

Overall, the tech startup ecosystem is dynamic and evolving, characterized by its agility, risk-taking culture, and potential for making a significant impact on industries and society at large. As startups innovate and disrupt traditional markets, they contribute to job creation, economic prosperity, and technological advancement, driving forward the global innovation landscape.

10.3 BANGLADESH'S JOURNEY TOWARD DIGITALIZATION: A REMARKABLE TRANSFORMATION

Bangladesh's journey toward digitalization is a testament to the country's resilience and adaptability, especially as a developing nation. Despite the challenges often associated with limited resources and infrastructure, Bangladesh has made significant strides in embracing digital technologies. The previous government's "Digital Bangladesh" initiative, launched in 2009 as part of Vision 2021, set the foundation for this transformation, aiming to harness the power of information and communication technology (ICT) to drive economic and social development. What is particularly impressive is how Bangladesh has managed to advance its digital agenda when some developed nations continue to face hurdles in implementing widespread digital adoption. The country's progress can be attributed to a combination of strong political will, an emerging tech-savvy population, and strategic investments in ICT infrastructure.

The adaptability of Bangladesh in embracing digitalization is highlighted by its flourishing startup ecosystem, which has thrived despite the challenges. Entrepreneurs like Shameem Ahsan, a tech pioneer and co-founder of eGeneration, have been at the forefront of this movement, driving innovation and contributing to the country's reputation as a rising tech hub in South Asia. These efforts collectively underscore Bangladesh's unique position as a developing country that has successfully navigated the complexities of digital transformation, setting an example for others in the Global South.

10.4 GROWING PRESENCE OF STARTUPS

Over the past few years, there has been a noticeable increase in the number of women participating in the tech startup ecosystem. More women are founding startups, taking on leadership roles, and contributing to technological innovations.

10.4.1 Notable Startups and Entrepreneurs

Shohoz.com: Shohoz's founder and managing director Maliha M. Quadir has built one of the leading service marketplaces in Bangladesh. The sole objective of Shohoz when they embarked on this adventure in 2014 was to simplify people's lives. Today, Bangladesh's most popular online ticket destination is Shohoz, a pioneer in the travel business. Being a tech-first business, they provide tech-driven solutions to address Bangladeshi people's daily problems. They allow them to select from hundreds of operators and routes, compare costs, provide the best offers, and take advantage of safeguards—all of which can be done in minutes and with a few phone taps.

Dhaka Cast: Dhaka Cast is a healthcare startup focused on supporting diabetic patients and is fully run by experienced and qualified doctors. Dr. Fahreen Hannan launched it in June 2019 with three other partners, including her husband, a doctor. It provides home sample collection, round-the-clock medical consultations, and online appointments with specialists and GPs. For testing, they collaborate with Turag Diabetic Center and Ibn Sina Hospital. Reports are sent via email or delivered, and doctors consult over video conversations. They also offer home visits by physicians and nurses and medication delivery. Evening hours are when specialists are accessible. During the pandemic, they provided care for about 5,000 patients, reaching a monthly peak of 1,300.

Praava Health: Founded by Sylvana Quader Sinha, Praava Health is a healthcare provider in Bangladesh known for its patient-centric approach and quality medical services. Established with a mission to deliver accessible and reliable healthcare, Praava Health emphasizes patient care, modern facilities, and a comprehensive range of medical services to meet the needs of the community (Sinha, 2024).

Maya: Founded by Ivy Huq Russell, Maya is a pioneering digital health startup in Bangladesh. It specializes in providing comprehensive health and wellness information and services, with a particular focus on women's health. Maya's platform aims to democratize healthcare by offering accessible telemedicine services and health-related content across digital platforms. The Maya app is accessible in four languages: Bangla, English, Urdu, and Sinhala; it serves approximately 7 million users globally each month (Jui, 2021).

Techmania: A notable example of female leadership in Bangladesh's technology sector is Taslima Miji, the CEO and founder of Techmania. The company provides a wide range of hardware-related products and services, including computer sales, assembly, networking solutions, and peripheral support, along with troubleshooting and technical assistance. Beyond its current operations, Techmania also nurtures new ideas and innovations that are still in the early, conceptual stage (Team, 2024).

Dcastalia: Dcastalia is a software development firm based in Bangladesh that was founded in 2009 to provide individualized, devoted, and customized services. After more than 10 years of business, they have established a solid reputation for dependability and credibility domestically and globally among clients from various industries regarding software development, website development, technical support, graphic design, and many other services (Dcastalia Ltd., n.d.). Sabila Enun is the Head of People and Communications at Dcastalia and one of the founders (Business Inspection BD, 2022).

Pulse Healthcare Services: The creator of Pulse Healthcare Services, founded in 2017, is Rubaba Dowla. It is a telemedicine platform that uses video conferences to link physicians and patients. With Pulse, people may safeguard others and themselves without waiting in line or visiting congested hospitals. With the help of Pulse Healthcare Services, patients may receive top-notch medical treatment from the comfort of their homes by connecting with doctors via a secure online video conferencing platform whenever and wherever they choose, provided they can access the internet on a laptop or smartphone. These online consultations are safe and practical, and the COVID-19 epidemic highlighted the importance of the platform (Pulse, n.d.).

Foodpanda Bangladesh: Ambareen Reza, the CEO of Foodpanda Bangladesh, is a significant figure in the tech sector of Bangladesh, particularly as a woman leader in an industry that is often male-dominated. Her journey reflects the growing presence and impact of women in technology and entrepreneurship in the country. Reza co-founded Foodpanda Bangladesh and has been instrumental in its rapid growth and success. Under her leadership, the company has become one of the leading online food delivery platforms in Bangladesh, demonstrating her strong business acumen and ability to navigate the complexities of the tech-driven food delivery industry. Her role as CEO positions her as a role model for other women aspiring to enter the tech sector, showing that with determination and innovative thinking, women can not only participate but also lead in technology-driven businesses. In the broader context of women in tech in Bangladesh, Reza's achievements are part of a larger movement where more women are breaking barriers in the tech and startup ecosystem. While the tech industry in Bangladesh has traditionally been dominated by men, women like Ambareen Reza are challenging the status quo and paving the way for greater gender diversity. Her leadership at Foodpanda Bangladesh exemplifies how women are increasingly taking on leadership roles, influencing the tech landscape, and inspiring the next generation of female entrepreneurs and tech professionals in the country.

Reflections: After acquiring expertise in creating artistic glass sculptures, Sabrina Islam established Reflections in Bangladesh in 1998, modernizing the country's architecture field. At first, she struggled to maintain Reflections because she lacked raw materials and experienced staff. However, with perseverance and hard effort, the business began to succeed more often. These days, it offers items including stained glass, etched colored glass, sandblasted glass, curved glass, and conventional glass overlay (Business Inspection BD, 2022).

10.5 SUPPORT AND INITIATIVES

The Government of Bangladesh, along with various NGOs, has initiated programs to support women entrepreneurs. These include funding, mentorship, and training programs aimed at enhancing their skills and providing them with the necessary resources. Programs like Women Techmakers Bangladesh and Accelerating Asia offer mentorship, funding, and networking opportunities. These women-focused incubators and accelerators are:

Women Techmakers Bangladesh: Rakhshanda Rukham founded Women Techmakers Bangladesh to empower women to pursue careers in STEM[1]. Recognized by Google, the organization has been supported by them since 2015. Women Techmakers is now one of the leading networks for IT students and professionals in Dhaka. Rakhshanda is passionate about using technology to empower women through community initiatives. She has helped formulate over 45 women-only tech programs, engaging over 4,300 participants to date. One such project, entitled "Begum," focuses on providing internet access to women in rural communities. "Begum" was selected as one of the top five women-led startups at SeedStars Dhaka and one of the top 25 in the Grameenphone accelerator. Rakhshanda has also campaigned to end sexual harassment at universities.

Accelerating Asia: While not exclusively for women, it supports many women-led startups in Bangladesh, offering mentorship, investment, and networking opportunities. Accelerating Asia was one of the first early-stage international venture capital firms to invest in Bangladesh, funding companies such as Amarlab, Dana, DoctorKoi.com, HandyMama, HelloTask, iFarmer, Loop, markopolo.ai, MedEasy, Nitex, PriyoShop, Romoni, Swap, Shuttle, Sohopathi, SupplyLine, and Zantrik.

10.6 COMMUNITY AND NETWORKING GROUPS

Platforms like Women in Digital and Girls in Tech Bangladesh foster community support, skill-building workshops, and networking events for women in startups. These networks provide opportunities for collaboration and knowledge sharing.

Women in Digital: Achia Nila, founder and CEO of Women in Digital (WID), established the organization in 2013 to promote women's and girls' education and empowerment through technology. The organization addresses the digital and gender divide in the IT sector by offering practical solutions and advocating for inclusive policies. Since its inception, WID has trained and empowered over 7,000 adolescent girls and women across Bangladesh, providing digital literacy and market-relevant IT skills such as online business development, digital marketing, and app development (Women in Digital, 2024). WID is the only women-led and women- and girls-focused organization in Bangladesh's IT sector, aiming to foster leadership and economic independence. Through its efforts, WID is challenging social and gender norms, striving for sustained, inter-generational impact.

Girls in Tech Bangladesh: This is a global non-profit that focuses on the engagement, education, and empowerment of women in technology.

10.7 SUCCESS STORIES AND IMPACT

Leadership and Innovation: Women-led startups in Bangladesh, such as Shohoz and Praava Health, have demonstrated leadership in sectors ranging from service marketplaces to healthcare. Their innovations address critical societal needs and contribute to economic growth.

Job Creation and Economic Empowerment: Women entrepreneurs create jobs and contribute to local economies, empowering other women through employment opportunities and leadership roles within their startups.

Role Models and Inspiration: Successful women entrepreneurs serve as role models, inspiring the next generation of women to pursue careers in entrepreneurship and STEM fields. Their visibility challenges stereotypes and encourages more inclusive business environments.

10.8 CHALLENGES AND BARRIERS

Despite the progress, women in Bangladesh's tech startup sphere face numerous challenges. Cultural norms often discourage women from pursuing careers in technology or entrepreneurship, leading to underrepresentation in these fields. Additionally, access to funding is a significant barrier, with female entrepreneurs often finding it difficult to secure investment compared to their male counterparts. The lack of mentorship and networking opportunities further compounds these issues, making it harder for women to scale their businesses.

Cultural and Societal Norms: Traditional gender roles and societal expectations often discourage women from pursuing careers in tech or entrepreneurship. Family support can be limited, and societal pressure to conform to traditional roles remains a significant barrier. In Bangladesh, women are typically discouraged from pursuing jobs in technology or entrepreneurship due to traditional gender roles and cultural expectations that place a higher priority on home obligations than professional aspirations. Women wishing to launch enterprises may experience lack of family support as a result of this social pressure. Strong familial ties and clearly defined gender roles are valued in the nation's culture, which is strongly impacted by its predominately Muslim population. This further restricts women's chances to pursue entrepreneurial endeavors. Rural areas continue to be more traditional, whereas urban areas are witnessing a shift with an increase in nuclear families and dual-income households. These areas present considerably more obstacles for women in terms of access to technology, education, and business prospects. Bangladesh is predominantly a Muslim country, with nearly 90% of its population adhering to Islam. Despite Islam being the majority religion, Bangladeshi culture has integrated many traditional beliefs and practices, emphasizing strong family bonds. In urban areas, about 70% of families are nuclear, with both parents typically working outside the home or engaged in income-generating activities (Hamadani and Tofail, 2014). However, traditional gender roles and societal expectations often discourage women from pursuing entrepreneurship. Additionally, family responsibilities and cultural barriers can impede their ability to start and grow businesses. Women's participation in the workforce is increasing, but significant challenges remain in achieving gender equality in business and entrepreneurship. Efforts to empower women through education, financial support, and policy changes are crucial for overcoming these obstacles and fostering a more inclusive entrepreneurial environment. All things considered, these social and cultural conventions pose serious obstacles for women working in digital startups.

Skill Gaps and Education: Although the number of women in STEM fields is increasing, there is still a gap in technical skills and higher education among women in Bangladesh. This can limit their confidence and preparedness to navigate the competitive startup landscape. In Bangladesh, women are pursuing STEM degrees, yet there is a huge talent gap. Women's access to high-quality education and training in STEM subjects is typically restricted, which makes them less competitive in the tech startup scene. Their confidence may be affected by their lack of technical knowledge and higher education, which could make them less equipped to succeed in the fiercely competitive startup world where technical know-how and creative problem solving are essential. Furthermore, because of their obligations to their families and societal norms, women usually have fewer options for lifelong learning and professional advancement. This may limit their potential in the tech startup space by making it more difficult for them to keep up with the industry's rapid improvements.

Access to Funding: Women entrepreneurs often face challenges in securing funding for their startups. Investors may have biases, and due to a lack of networking opportunities and limited financial independence, women entrepreneurs in Bangladesh face major obstacles when trying to secure financing. Because of their preconceived notions about gender roles in the workplace, investors frequently retain unconscious biases that cause them to discount the abilities of women and favor male-led firms. Furthermore, women entrepreneurs generally do not have access to the vast networks that male entrepreneurs do, networks that are essential for obtaining capital and learning about the market. These networks provide women with access to financial support, collaborations, and mentorship that they frequently do not receive otherwise. It's also possible for Bangladeshi women to lack financial independence, which makes it harder for them to save up the money needed to launch their businesses. Their ability to autonomously make important financial decisions may also be limited by cultural conventions, which would further impede their business endeavors.

Lack of Mentorship and Role Models: There are fewer female role models and mentors in the tech industry, which can make it challenging for aspiring women entrepreneurs to find guidance and inspiration. Female role models are scarce in the technology sector of Bangladesh, which is a major obstacle for aspiring female entrepreneurs. In the software industry, there are few prominent female role models, which reduces newcomers' access to advice and motivation, feeding the cycle of gender inequality in the sector. A major contributor to this problem is the lack of mentorship programs available to the general public. Most women have difficulty finding mentors who can provide them with guidance, encouragement, and professional connections. To launch and expand a tech company, women need an effective mentor, but many lack access to these valuable resources. Two obstacles prevent women from becoming successful entrepreneurs in the digital space: lack of role models and lack of mentorship.

Work-Life Balance: Since women in Bangladesh are usually the primary caretakers, they frequently shoulder heavy family duties. Having to handle both managing domestic responsibilities and growing a firm can cause fatigue and lower productivity. It is more difficult for women to completely devote themselves to their entrepreneurial endeavors due to a lack of supportive legislation and flexible work settings. Without sufficient assistance, juggling the demands of a startup along with obligations to one's family can make it more difficult for founders to successfully grow their companies. Enabling women to thrive in the digital startup sector requires putting in place rules that promote working mothers and flexible work arrangements.

Legal and Policy Barriers: Significant regulatory barriers, such as convoluted bureaucratic procedures and a dearth of gender-sensitive laws, face female entrepreneurs in Bangladesh. These difficulties may hinder their development and discourage them from pursuing a career in IT startups. Furthermore, there is a lack of institutional and governmental support expressly designed for women in digital entrepreneurship, which limits access to resources. Creating a friendly atmosphere requires policies that favor gender equality and offer necessary resources, including money, mentorship, and training. By improving these regulations and offering focused assistance, these obstacles can be removed, allowing more women to succeed in the IT sector and contributing to a more diverse entrepreneurial environment.

Cultural Perceptions of Failure: In Bangladesh, there is a strong shame associated with failing businesses, especially for women. This societal view deters women from pursuing entrepreneurial endeavors since the consequences of failure can be severe on both a social and a personal level. Social expectations, which frequently subject women to more scrutiny and pressure to fit into traditional roles, amplify the fear of failure. As a result, women are not willing to take risks, which limits their capacity for innovating and ensuring company growth. Their lack of willingness to take calculated risks restricts their ability to be entrepreneurs, as these kinds of risks are essential to innovation and business growth. To address this, societal changes that accept failure as a necessary part of the entrepreneurial process and support networks that inspire women to continue and persevere in their business ventures in the face of any obstacles are required.

10.9 THE FUTURE AHEAD

Looking forward, the future for women in Bangladesh's tech startup sphere appears promising. As awareness grows and more support structures are put in place, more women will likely enter the field, further enriching the ecosystem. However, continued efforts are needed to address the persistent challenges and ensure that women entrepreneurs can thrive. Let us discuss potential future trends and the steps needed to create a more inclusive and equitable startup environment. One of the most significant drivers of future growth for women in tech startups is the increasing digital literacy across Bangladesh. With the expansion of internet access and the proliferation of smartphones, more women, particularly in rural areas, are gaining access to digital tools and resources. This increased access is expected to empower more women to explore entrepreneurial opportunities in the tech sector, as they can now leverage technology to create and scale businesses from virtually anywhere. Government initiatives aimed at promoting digital literacy, such as the "Digital Bangladesh" vision, play a critical role in this transformation, ensuring that women are not left behind in the digital economy. The future success of women in tech startups will also be bolstered by the expansion of support networks and mentorship programs. As more women enter the tech startup space, the need for peer networks, mentorship, and role models will grow. Organizations such as Women in Digital, Bangladesh Women in Technology (BWIT), and She Loves Tech are already laying the groundwork by providing platforms for networking, skill development, and mentorship. These organizations are likely to expand their reach and influence, helping more women navigate the challenges of entrepreneurship and build successful tech ventures (The Daily Star, 2023). Additionally, collaborations between local and international organizations can further enhance the support available to women entrepreneurs, providing them with global perspectives and opportunities. Access to funding remains one of the most significant challenges for women in tech startups. However, there is growing recognition among investors of the need for gender-inclusive investment strategies. Research shows that diverse teams and women-led businesses often outperform their counterparts in terms of innovation and profitability. This realization is expected to lead to the development of more gender-focused venture capital funds and angel investor networks in Bangladesh. Initiatives like the Bangladesh Angel Network and international investors increasingly focusing on gender diversity could play a pivotal role in ensuring that women entrepreneurs have better access to the capital they need to grow their businesses. The role of policy and regulatory frameworks in shaping the future of women in tech startups cannot be overstated. The Bangladeshi government has shown a commitment to promoting entrepreneurship through various policies, but more targeted efforts are needed to support women entrepreneurs specifically. Future policies could include tax incentives for women-led startups, grants for research and development, and initiatives to reduce the regulatory burdens that disproportionately affect female entrepreneurs. The government's ongoing focus on digital transformation, coupled with a gender-sensitive approach to policy-making, will be crucial in fostering an environment where women can thrive in the tech startup ecosystem (Alom et al., 2025). As more women enter the tech startup sphere, we can expect to see an increase in innovations that address issues particularly relevant to women. For example, health tech, education tech, and financial tech solutions tailored to the needs of women could become more prominent. Women entrepreneurs bring unique perspectives and insights to these industries, often identifying gaps in the market that have been overlooked by their male counterparts. The rise of women-led innovations not only diversifies the tech landscape but also ensures that technology serves a broader range of societal needs, contributing to more inclusive growth.

10.10 CONCLUSION

Despite challenges, the future looks promising for women in the tech startup ecosystem in Bangladesh. With continued support from stakeholders, including governments, investors, and community organizations, women entrepreneurs can overcome barriers and thrive in the dynamic startup ecosystem. Initiatives focused on education, mentorship, and access to funding will be critical in nurturing their talents and fostering a more inclusive entrepreneurial landscape. The tech startup ecosystem in Bangladesh is evolving, and women's contributions are increasingly recognized as vital to its growth and success. With persistent efforts to address the challenges, the landscape will continue to become more inclusive and supportive of women in technology. In conclusion, women are becoming an integral part of the tech startup ecosystem in Bangladesh, driving innovation and contributing to the country's digital transformation. Their participation is essential not only for achieving gender equity but also for fostering economic growth. By continuing to break down barriers and provide support, Bangladesh can ensure that women play an even more significant role in the tech industry, paving the way for sustainable growth and a more inclusive economy.

NOTE

1 STEM encompasses science, technology, engineering, and mathematics, providing an integrated approach to learning and development.

REFERENCES

Alom, K., Rahman, M. Z., Khan, A. I., Akbar, D., Hossain, M. M., Ali, M. A., & Mallick, A. (2025). Digital finance leads women entrepreneurship and poverty mitigation for sustainable development in Bangladesh. *Journal of Innovation and Entrepreneurship*, *14*(1), 34.

Business Inspection BD. (2022, September 26). *Top 10 women entrepreneurs in Bangladesh.* Available at https://businessinspection.com.bd/top-women-entrepreneurs-in-bd/

The Daily Star. (2023). Digital Bangladesh: A story of transformation, resilience, and sustainability. Available at www.thedailystar.net/opinion/views/news/digital-banglad esh-storytransformation-resilience-and-sustainability-3222626

Dcastalia Ltd. (n.d.). *Dcastalia | Software development company in Bangladesh*. Available at https://dcastalia.com/

Hamadani, J.D., & Tofail, F. (2014). Childrearing, motherhood and fatherhood in Bangladeshi culture. In: Selin, H. (ed) *Parenting across cultures*. Science across cultures: The history of non-Western science, Vol 7. Springer, Dordrecht. https://doi.org/10.1007/978-94-007-7503-9_10

Jui, U. M. (2021). Maya: An AI-powered start-up that plans to become the go-to name in healthcare. *The Business Standard*. Available at www.tbsnews.net/features/panorama/maya-ai-powered-start-plans-become-go-name-healthcare-314215

Pulse. (n.d.). *Pulse Healthcare Services – for your wellness*. Pulse. Available at www.mypulse.com.bd/

Shareef, M. A., Dwivedi, Y. K., Hosain, M. S., Giannakis, M., & Ahmed, J. U. (2024). Entrepreneurial ecosystem of startups in Bangladesh: Mechanism of balanced coexistence. *International Journal of Entrepreneurial Behavior & Research*, *30*(5), 1314–1346.

Sinha, S. Q. (2024). The CEO and founder of Praava Health on reimagining care in an emerging market. Available at https://hbr.org/2024/05/the-ceo-and-founder-of-praava-health-on-reimagining-care-in-an-emerging-market

Team, F. S. (2024, July 14). *Face to face with Taslima Miji, founder and CEO, Techmania – Future Startup*. Future Startup. Available at https://futurestartup.com/2013/03/08/face-to-face-with-taslima-miji-founder-and-ceo-techmania/

Women in Digital. (2024). About us – Empowering Inspiring Accessible. Available at https://womenindigital.net/about-us

Index

For Product Safety Concerns and Information please contact our
EU representative GPSR@taylorandfrancis.com Taylor & Francis
Verlag GmbH, Kaufingerstraße 24, 80331 München, Germany

W